GABE OPPENHEIM

Borrowed Time

A History of Espionage, IP Theft, and the Global War for Horology

First published by Solicitude 2026

First edition

ISBN: 979-8-234-05464-7

This book was professionally typeset on Reedsy.
Find out more at reedsy.com

For Lily, who was on the clock before she ever emerged; who spontaneously unspooled from her coiled place ahead of time; and who, subject now to the rote scheduling of grown-ups, we hope will retain nevertheless that free-spirited rhythm.

The clock turns time from a process of nature into a commodity that can be measured and bought and sold like soap or sultanas.

— George Woodcock

Contents

Preface

The greatest feeling we're afforded by time is that of propitious timing itself – when in a moment, or immediately thereafter, we perceive our salvation in a matter of random chronological good fortune: We got the gig, we made the sale, we found a spouse for having been in the right place at just the right time. If we hadn't glided onto that Subway car or into that elevator. If we'd grabbed a drink at different bar on a different night.

The sliding doors story.

This, of course, is a contrivance: It's perhaps the case that no matter the timing, we always would have found such reward – or, put another way, inescapable were our destinies. We had but one job we were fated to work, but one soulmate we were penciled in to meet. Or we're not discrete beings at all – we're all of us composed of the same atoms – there is not a cosmic difference whether "I" send that tweet or "you" do. Such divisions themselves are artificial (and, tautologically, *divisive*).

Or maybe we fail to see the chaos: the elevator we didn't get on had a better boss at a better company there for the pitching. Nine other potential spouses at nine other bars awaited a hero's entrance – but we never made it to those watering holes. We missed out on potentially more enriching life partners.

How can we account for the opportunities of which we will always remain unaware?

Argument: Time is merely our processing of the chaos into which we're daily thrust. It's the sausage casing for unsightly animal bits – the gross organs we couldn't possibly stomach without their being enveloped.

So wrote Albert Einstein, with co-author Leopold Infeld, in a 1938 book:

Physical concepts are free creations of the human mind, and are not, however it may seem, uniquely determined by the external world. In our endeavor to understand reality we are somewhat like a man trying to understand the mechanism of a closed watch. He sees the face and the moving hands, even hears its ticking, but he has no way of opening the case. If he is ingenious he may form some picture of a mechanism which could be responsible for all the things he observes, but he may never be quite sure his picture is the only one which could explain his observations. He will never be able to compare his picture with the real mechanism and he cannot even imagine the possibility or the meaning of such a comparison.[1]

The universe to us is the watch mechanism we can never grasp – the timepiece whose back is permanently sealed.

What Einstein failed to add is the tremendous satisfaction our intuition of the watch gears can provide – that the story of physics is perhaps no less delectable for being still incomplete, inconclusive, infinitely fuzzy even as we bore further in.

There is gratification in endless pursuits – in scrutiny perpetual. In narration where precise explication eludes us.

Enter Edgar Fogel Magnin, the Reform rabbi to the stars in the Golden Age of Hollywood – the man who, leading the Wilshire Boulevard Temple, both married MGM mogul Irving Thalberg to Norma Shearer and read the Kaddish at Thalberg's funeral not long after when pneumonia overtook him (besides later leading the final rites at the services for Jack Benny and Eddie Cantor).

Born to department store magnates in San Francisco, Rabbi Magnin later befriends the future First Lady Pat Nixon when she is a student at USC, which earns him a place in Tricky Dick Nixon's second inauguration ceremony – at a Sunday morning kaffeeklatsch of three faith leaders (Rabbi Magnin will tell Nixon the odd story of three magicians who made a tiger appear before a simpleton; the magic works but the tiger consumes his conjurers – meanwhile the simpleton, having climbed a tree in fear, escapes;

the moral was supposedly that common sense can trump long and esoteric study – which feels a compliment most backhanded, if it's a compliment to Nixon at all).

But 42 years earlier, Rabbi Magnin glad-hands a very different celebrity: He leads his congregation, the entire Jewish contingent of his city in point of fact, in welcoming Albert and Elsa Einstein to La-La Land at a reception at The Ambassador Hotel.

Enter the decade-old hotel's Italianate lobby from Wilshire — the floors covered in exotic rugs, the walls adorned with lush fabrics, crystal chandeliers poised above, a large stone fireplace to the side. Descend a particular staircase and you're in the Cocoanut Grove club. Just two months earlier, the producer Myron Selznick got into a fistfight with a less notable name seemingly over the attention of Jean Harlow.[2]

But tonight, February 17, 1931, we enter an enormous Ballroom for a less pugilistic affair. Rabbi Magnin and his congregants gift Einstein a forest — one that has been planted in his honor in Jerusalem – a land very much at the center of this congregational orbit. Einstein wishes for peace in the land.

And then there's another gift, for the man who understands our universe to be never fully understandable – who deems it a watch mechanism unseen, a gear train elusive, levers and springs that can only be imagined:

A 14k gold tonneau-shaped wristwatch – manufactured by, and branded with the name of, Longines. Its hairspring owes a good deal of its precision to the so-called Breguet overcoil. It will be the piece by which Einstein measures time, in a certain sense, for the rest of his life.

But neither the Rabbi who presents that offering nor the physicist who so graciously receives it truly know its provenance — how the piece soon to take a semi-permanent place on Einstein's wrist bears the name of an outfit that only decades earlier took steps that undermined America's greatest timekeeping company (back when the US watch industry arguably outpaced that of the Swiss).

Einstein walks around in a Longines unaware that company's most legendary leader and engineer engaged in an industrial espionage operation in America — a corporate mimicry of timekeeping manufacturing that by other opportunistic actors (say, superclone Chinese watch creators) continues wildly unabated today.

* * *

It starts with a question — one you might not think to ask at first but can't really shake once you have: How did a tiny nation — however propitious its geography, educational system, coo-coo clock-crazy, war-avoidant culture (shout-out to Welles in "The Third Man") — retain *over centuries* a dominant position atop an ever-more globalized industry?

In simpler terms : How and why is "Swiss Made" still an imprimatur of luxury on watch dials? How are Swiss timepieces even still a thing?

Hell, even Toblerone is technically no longer a Swiss product; three years ago, its maker, Mondelez, had to forfeit *by law* its proclaimed Swiss provenance when it opened up a factory line in Slovakia.

The Swiss are known (beyond chocolate) for their banking prowess — and yet in the year Mondelez lost its official status as national product, the number two bank in the country, Credit Suisse, crashed, forcing the only bigger bank in the cantons, UBS, to buy the business (and even then, the government had to inject UBS with $100 billion of liquidity).[3]

Other Swiss companies — such as the formerly private flavor and fragrance firm Firmenich, whose former CEO Ilaria Resta now leads high-end watch purveyor Audemars Piguet — have merged with large foreign corporations (Firmenich joined with public Dutch chemicals company DSM the same year Toblerone lost its origin rights and Credit Suisse collapsed.)

None of which is to say that Swiss watchmaking has always been a healthy domestic industry free of financial issues and impervious to foreign M&A schemes. It's to point out the great oddity of its survival at all, let alone its preeminence (even if some of those brands still hunkered down in the lush Swiss valleys today are now technically owned by conglomerates such as Richemont, LVMH and Swatch).

Because many other countries had their own artisanal and industrial watchmaking bases from which to draw talent and take away Helvetic market share: England and Germany and America and Japan have all brimmed at various points in the last three centuries with mechanical timekeeping innovators and visionary factory-floor engineers. All these countries have cultivated the talent, the technology, the culture to fight for that top role.

Why — contrary to the entropic take of Yeats and then an allusive Didion — did the center actually hold? How have the Swiss retained their horological supremacy? Has it been by hook…or by crook?

That is the first query that animates the following text. The second: If the Swiss did employ at times questionable methods to retain their watchmaking lead, did those sins ever catch up to them? Were they ever made to pay a karmic price?

Was there a market that did to the Swiss what the Swiss had long done to others?

***

A disclaimer: Industrial espionage is not necessarily illegal — and the laws of the US and Switzerland have obviously evolved since the 19th-century (*and* the 20th). The terms "thievery," "heisting," "espionage" and all their synonyms denoting a certain kind of *taking* — they are employed solely in the service of telling a story entirely truthful about a transference of knowledge from one fascinating figure to another. To convey facts and let the reader judge them. Aside from those figures in the book actually convicted of crimes, no epithet or description herein is meant to indicate a violation of law.

Of course, a measure may be immoral while not quite being illegal.

Introduction

In the smog that hangs over the Yalu River, two spans emerge from the northwest, heading southeast toward another land – but only one actually makes it, reaches the far banks.

That is the Sino-Korean Friendship Bridge, and as its aggressively cheerful name suggests, it connects China and North Korea – more specifically, the coastal Chinese province of Liaoning with the North Pyongan section of North Korea.

Here, the great communist power and its hermit kingdom adjunct meet – here is the border between a society controlled by the state and one even more unyieldingly squeezed by that apparatus.

But there is another bridge here, one that runs parallel to the 3087-foot Sino-Korean. They call it the Broken Bridge, because it juts only slightly more than halfway toward North Korea, its unused pilings, like totemic guardians, visible above the turbid water.

Here, you can walk toward North Korea as if you were planning entry, before the roadway runs out – before you are halted at the edge, the roadway simply gone.

No, the Broken Bridge does not reach the other side – does not even seem to try – the construction calls to mind the opposite of one's grasp exceeding his reach.

This structure was partly bombed by the US during the Korean War 75 years ago and then, though its rebuilding on the Korean side would have been relatively simple, torn down altogether and scrapped on the North Korean side of the river – a structure abandoned to history, frozen in time, the resultant void evocative of the entire country's radio silence.

From the Broken Bridge, you can gaze out at a land forbidding and bleak, meditate on the strangely arbitrary nature of lines on a map, of nation-state discontinuity, but that's it – the structure prohibits real exploration.

If only for this reason alone, for its frontier nature, does the Chinese city from which these bridges emerge merit scrutiny – Dandong, in the aforementioned province of Liaoning.

But Dandong is not just the last stop before geopolitical abyss.

It is a hub once hulking and deindustrialized – a city not unlike Pittsburgh or Youngstown, Baltimore or Detroit, although Dandong's golden days of production date from the mid-1980s to the turn of the millennium. But in that golden window, Dandong hummed.

Its chemical plant produced a major herbal toothpaste – Kang Chi Ling. Its automotive firm, once just a maker of axles, produced popular Huanghai busses and Shuguang SUVs. And then there were its watches – the Liaoning Watch Factory, founded in 1957 by the national government as one of the first four watchmaking enterprises backed by the Communist state, was by 1986 producing 3.85 million watches per year (second-most among all Chinese horological concerns). It employed 12,000 people, its profit 120 million yuan.[4]

And then China opened up to global brands and pivoted its economy toward pursuits technological. And as a result, the center of manufacturing drifted south – toward Shenzhen, and more broadly, Guangdong, the mainland region just opposite Hong Kong whose lucrative role in the Internet economy meant it could dictate production in less-connected industries, as well. Dandong was caught by surprise: a city unprepared for real competition, unaware a game was even afoot, suddenly found itself losing that contest.

Hollowed out and embarrassed. Poorer.

But return to the heyday, to 1986. You are on the Liaoning factory floor. Small gears, pinions, arbors abound. Hundreds of your work friends — maybe work lovers — crouch over these tiny mechanical parts in an endless row of stations besides you, magnification loupes pulled up above their eyebrows and secured there by headbands, like secular, repurposed phylacteries. The room is practically silent given the intensity of the work, the concentration required. And yet, there is the sound of a humming fan, most likely, and more than that, there's the din of silence itself — that inaudible but palpable thrum of a giant collective workshop, the sound of shared purpose.

And then return just 20 years later, when the stillness on the factory floor is only that — a void. When dust has settled on long-unused desks, machinery, workshops.

Dandong: The toothpaste-maker – Kang Chi Ling's parent company – was sold four times. Then the Liaoning watch factory was acquired in 2008 – swallowed by a larger concern with no particular sentiment for Dandong's horological legacy. The Shuguang auto manufacturer, the one with the Mitsubishi engines and muscular names – Challenger, Conqueror – relocated its headquarters and R&D center to Beijing in 2020 – to reboot itself far away from the Yalu River.

And yet, there is also a turnaround story – and a third bridge. This one is south of the other two, very slightly downriver – the New Yalu River Bridge, whose spiraling cables extending from the two towers lyre-like, the taut strings of an strummed harp, make it seem utterly modern (Western, even). The project was basically completed in 2014, although the bridge hasn't opened due to delays on the North Korean side in the erection of customs facilities – small measures North Korea could take if it truly desired to open up, ease the flow of people and goods.

But China has endorsed a revival on its bank, in the Zhenxing District of Dandong, a city now of 2.2 million people: Once off the new bridge's exit ramp, you need walk just one and half blocks west along Guomen Avenue, and then one block south on Yanghe Street, for proof. There, on the western side of the street lies a facility for Huanghai Automobile, whose electric busses are now exported to South Korea, among other markets.

And on the Eastern side of the street, the beautifully manicured campus, replete with a hedge maze and a full-length outdoor basketball court, of the Liaoning Watch Industrial Park, the 100,000-square-meter home since its opening in 2011 of the renamed Liaoning Peacock Watch company, which focuses on making mechanical movements for other brands' timepieces (the engines of non-quartz, non-digital watches). And a few select very expensive watches under the Peacock brand – watches with dual tourbillons that retail for several thousand dollars each.

In 2012 it fashioned 100,000 movements in total. It now produces between 1 and 1.5 million of them annually. By 2020, its profit after tax was about $7.22 million US dollars.

That still puts Peacock a distant second in Chinese movement-making behind Sea-Gull – a company based in Tianjin, a city of 13.9 million – that produces between 3 and 5 million movements.

And yet Dandong is alive once more.

Not that the revival comes without a darker underbelly element.

For my book research takes me down both the legitimate rabbit hole of horology and its illegitimate shoot, too. And I find out the best clone Rolex, Patek Philippe and Audemars Piguet movements presently produced are constructed by "DDF" – also known as "Dandong Factory."

When I began sourcing black market pieces, I suspected the movements were produced in the shadows of the city's reborn industry – that DDF was an underground offshoot of the reputable Liaoning Peacock brand, an operation that thrived precisely because the city's proper watchmaker had begun to – but nothing further. A small remora suction-cupped to the top side of a shark.

And then a high-placed Swiss source told me that *Liaoning Peacock Watch Factory* itself was cloning movements for money. It had begun a half-decade earlier with a Rolex 4130 chronograph mechanism (the engine that drives the famous Rolex Daytona, the favored watch of Paul Newman once upon a time). And then authorities busted Peacock in 2021, and it made efforts to reemerge as an aboveboard entity — a manufacturer of original watch engines alone.

In reality, it began selling a *modified and unbranded* (also known as "sterile") version of the Rolex 4130 movement – whose general patent protection had mostly elapsed the prior year, in 2020 – called the SL-4801, which can be found in any number of legitimate chronographs today. This was a move very much grey — not in contravention of the law but not in its spirit either.

And then Peacock began producing clones of Patek and AP movements.

My superclone Patek Aquanaut is powered by a DDF 324 clone caliber, which runs like a dream. My Clean Plus Factory-produced titanium Rolex Yacht-Master and my Victory Factory-made Rolex Sea-Dweller are powered by buttery smooth DD3235 movements. The DD stands for "Dandong," home city of Peacock, by all accounts.

What exactly compelled Liaoning Peacock to continue crafting illicit calibers, years after that first bust, its shameful exposure?

Consider the three bridges emanating from Dandong across the river – one old and insufficient – a way of doing things perhaps no longer viable. One new and unopened – a successful venture that seems presently elusive but could work out if only one day it were given proper attention. And the one that stops halfway – that will never reach the other side – a bridge that brings you up to the very border of another place, but no further – a Gatsby-esque construction from which Daisy's light can only be seen, her world an impossible reach.

Peacock Watch Factory – it may have attempted to escape the underworld murk. Perhaps it had intended at one point truly to go honest. Abandon the counterfeit trade. But it hadn't, a source told me. Perhaps it just couldn't make it any other way.

We are all drawn to those things we do best.

My legitimate Peacock factory watches.

1

America's Birthday, 1876

It begins at the end.

A grandfather of about 70 years – a survivor of camps and death marches, of ghastly solutions enabled by the meticulous measurement of time, by chronology cold abetted – enters a deli in a New Jersey suburb just over the George Washington Bridge.

Or perhaps it was the pizza place or the chicken nugget shop.

His hair is white, thicker in places than it has any right to be (as is his waist, given his history of heart trouble); his smile's the irrepressible grin of a man exultant in the very notion of descendance – that these 10-year-old and 7-year-old boys (give or take – the memory fails some) emerged from his line.

His survival was our existence. Our existence is his jubilation. So long as we eat.

Who knows what we talk about with Grandpa while we chew – only that at some point, he hands my nearly-pubescent self a gold watch. I take it in my hand, reviewing several congruent notes: This piece is too insubstantial, thin, oddly reflective of the room's lighting, the band expandable as if its links are connected by elastic, to be real gold.

More importantly, I have scanned my mother's face, and she seems fine. Had he ever tried to gift me real gold, she would have snatched the item and insist he return it. Anyway, Grandpa's not the type to lovebomb his *kinder* with luxury items (unless you consider the Nerf guns their mom has banned such an indulgence – and they really were).

My grandpa says something about it being a good thing for me – but in my recollection, whatever he says does not compare with his consistent and regular praise for Seiko. *Now, you know who makes a* good *watch? Seiko.*

It all feels like a $15 purchase at CVS, and I'm content with that – I never sought such a wristwatch and its transfer has ruffled no family feathers, broken no parent-grandparent ground rules (around this period I will begin wearing a used red-and-black G-Shock gifted me for my birthday by a wealthy South African family friend who no longer has use for it, but that must not have happened yet).

I gaze at the glossy, golden dial (maybe it was more matte, with a subtler sheen than the bracelet) and sound in my head the printed name of the corporation responsible for its production:

Waltham.

At the time, I had no way of knowing that this printed name belonged not just to a company but to a lost capital of watchmaking – a place whose methods would be studied, coveted and, in the end, stolen. Nor could I have guessed that the history of watches, so often sold as a triumph of Swiss purity and precision, would reveal itself to be crowded from the start with frauds, forgeries, impostures and men taking notes where they had no right to be.

* * *

The watch world is thick with deceit – with counterfeit products, doctored provenance papers, mendacious (or, at the very least, unctuous) salespeople. The trade learned early how much money there is in borrowed legitimacy. Imposture.

Paris

Tuesday, September 12, 1893

Thunder – clapping at intervals irregular – nature's unnerving, clangorous chimes – announces an overbearing rain – a downpour that drenches the grape fields of Bordeaux and Perpignan and prevents workers from harvesting their autumn crop. In more urban areas, sewers swiftly overflow.

And a railway is submerged – one which runs to Perpignan from Paris, stalling one express train by a full four hours. The delay gives those stifled occupants more time to gossip about the great criminal scandal currently unfolding in their capital city: Have you heard – the gendarmerie have arrested three men – a Swiss watchmaker, a local craftsman of sepiolite (the mineral otherwise known by its German name "meerschaum") smoking pipes and a Polish man – one Jacobowski. All seem to be Jews, just by virtue of their names. And the crime – do you not know the details? They have persuaded a perfumer – one Monsieur Bertrand – a man who also manages an underground betting club – that they'd located in Germany a supply works that could provide at incredibly low cost an amber ingredient for the man's fragrances. And not just that, Monsieur Schwob, the ringleader, has taken from the perfumer a pearl necklace on consignment to sell. And yet all of it was fraudulent, fabricated, conniving. There never was an amber supplier requiring payment nor did the arrested men sell the necklace. All told, they've pocketed on the false pretenses £69,200 or 1,700,000 francs.

As the train reached its final destination in Perpignan, a Mediterranean town near the border of Spain dominated by the gothic, medieval palaces of Mallorcan kings, the chattering class probably put down the tales they'd just told to hyperbole – such a bizarre series of deceptions was probably less sensational than they'd heard. Or maybe in keeping with a certain anti-Semitic character, they ascribed these doings to the regular habit of Jews to trick and steal.

But even if they'd believed every brazen bit of the latest news – all of which was, in fact, accurate, by every reported account of the day from European and US correspondents both – they'd have unknowingly underestimated the bizarreness of the proceedings. Or how deeply the duplicity at their core touched upon not a religious minority at the fringe of European high society – but uppermost echelons of their old-world communes.

See, Armand Schwob was no mere Parisian watch retailer on the Boulevard Bonne Nouvelle. He'd once occupied an office, too, in the Jura region of

Switzerland – 14 Rue Leopold-Robert.

And it was from that seemingly high-class Swiss address that Schwob retailed, with the ostensible full authorization of its makers, the pocket watches of the *haute horlogerie* outfit known today as Patek Philippe.

Of course, that authority was not what it seemed. See, Schwob – he is the obviously criminal antihero whose brush with the established names in horology yields their own better-hidden misbehavior. Through Schwob we reach Patek. Through Patek Longines.

* * *

Schwob leads us backward. Eight years before his arrest, his company surfaces at a world's fair in Belgium – and it is there, in the records of that exhibition, that a thread emerges connecting a small-time swindler to the most powerful name in Swiss horology.

Antwerp

February 5, 1885

The thing about time is that, if you continue rolling it backward, you find at its beginning something uneasy, impure, apocryphal. What story hasn't by time's forward march found itself garbled and confused by misinterpreted murmuring across generations?

Perhaps that's the source of our original sin – our uncertainty about what came before, at our true source, due to neglectful transmission.

Or maybe we are afraid to recall the past because we can't have always been pious. Because at least once in our lives, we did wrong – perhaps an act that doubled and tripled, compounded by circumstance and silence – one whose metastasis, were we to recall it, blots out ever more, in the external world, but more so in the conscience.

Eight years before the arrest of Armand Schwob, his watch firm is represented by a stall at the 1885 Universal Exposition in Antwerp, Belgium. The late 19th-century has been filled with such year-long global gatherings, all aiming to promulgate the very best technical and political ideas of the industrialized west to nations near and far – and the exotic traditions of the Oriental, Southern and Island nations to the grand-touring class for their amusement. The ultimate stage of the Enlightenment: the selfless sharing of its fruits, not without condescension.

On February 5, 1885, King Leopold of Belgium seizes a rich vein of land in the center of the African continent – the Congo – labels it farcically "Congo Free State" and assumes its possession not as the colony of the country he rules but as his own personal landmass – the largest and most brutally-acquired backyard of its time.

Three months later, the Universal Exposition, staged across 54 acres, opens its proverbial doors, and King Leopold invites to the fair 12 prominent Congolese men, including King Massala, so they may undertake a study of civilized government. The fair is also divided into various industrial sections – businesses should surely learn from each other as much as national leaders do. And what better way to encourage innovation than to pit enterprise against enterprise, such that sections have a ruling jury entitled to hand out prizes?

This includes an asbestos firm competition – won by Irwin, Hopper and Company – which had taken a medal at a universal expo in Canada two years earlier as well. Other more enticing wares on display include wine, chocolate, and cigars. A factory from Geneva proudly proclaims its production of butter, candles and soap, while an optician named Lugano travels all the way from Havana. Plus French sculpture of Sevres and blown glassware from Murano.

Let the flowing wine — its mouthfeel perhaps oaky, slightly fruity, intimating tobacco in its lingering beads— cascade down your throat. Taste, too, the tannins of bitter cacao — astringent, acidic like unroasted coffee beans — as curlicues of nearby puffers' rich stogie smoke — leathery and earthen at first, then vanillic, creamy — climb up your nostrils.

Now would you like a fine pair of spectacles so that you might survey the scene more sharply — obtain a finer view of its old-world exchanges, all of it like someone had crossed a bazaar with a rare-book library's members-only lounge?

And yet the sensory showcases – not to mention the teeming crowds, who'll ultimately make 3.5 million discrete visits – leave at least The New

York Times reviewer nonplussed.

"Glad to have gone there," he writes upon his return, "and still gladder to have come back in safety with all my luggage." He has found the entirely-booked hotels unhelpful and unyielding, and the claims about the exhibits "devices of puffery" voiced by "charmers."[5]

Yet the category of timekeeping must have been exception to this judgment – its field rich with brilliantly-devised chronometers, conceived by world-renowned minds – all of them being judged by a jury chaired by Adrien Philippe – of the eponymous firm known later as "Patek Philippe" but called now "Patek, Philippe & Cie".

On display are watches fabricated by Ernest Francillon of Longines, Charles Tissot and Louis-Ulysse Chopard – names that won't soon be forgotten in the trade – or the wider world.

And then, on Thursday, July 9, the aforementioned chair of the jury, Adrien Philippe, comes upon a most unusual exhibit.

> *One of the most impressive displays in the hall bore the name Armand Schwob & Frère. These Parisian brothers worked with craftsmen throughout the Swiss Jura region, representing their best products through an office in La Chaux-de-Fonds. Unlike many in the city and region, Schwob produced a wide range of watches in many different styles and even bought the products of others for wholesale. The firm boasted that its annual sales reached four million francs, an unbelievably large amount in the 1880's!*
>
> *When the jury arrived at Schwob's display case, they were met by a representative of the factory with a tray of pre-selected watches to examine at a central table. This was highly unusual – the jury typically gathered around the display case and picked the specimens themselves to evaluate. As noted in a contemporary report, "this eagerness seemed to us to include a hidden irregularity; consequently, we asked him not to trouble himself so much, preferring to select items ourselves at random."*[6]

And then, Philippe senses something awry, proficient as he is in taking not just the measure of time but of a man.

His jury, after reviewing suitably finished watches, with properly advanced mechanical movements, notices lesser pieces – shoddy and rough, of an inferior quality, in Armand Schwob's booth. These Philippe and his disciples desire to see most of all – the jury never expecting, their indistinction be damned, to encounter a line engraved on one interior with an almost ghostly quality:

"Pateck & Cie."

The watches also all bear the inscription of "Geneve," which seems an impossible source of said timekeepers, considering Armand's office is located in Neuchâtel, and he has claimed to acquire works from parlors across the Jura mountains (of which Geneva is not a part).

But far more remarkable, spectral and unseemly, is the inscribed company name.

Because the man responsible ultimately for the production all Patek Philippe watches (Patek himself having died 8 years earlier) is standing opposite the watch, delegated to adjudge its quality – Adrien Philippe himself.

He has most assuredly never seen this piece before, that surname of his late partner spliced (rather fiendishly, considering the man is dead) with an added "c."

Pateck.

It is perhaps the first case of luxury counterfeit watch production in modern history – and as such, the Swiss jury has no protocol. But they are honest men, hardworking, devoted to craft above commerce – and surely, they will report what seems an evident crime to the authorities.

But this is not the case – at least not initially, although eventually Mr. Philippe will take Armand the trickster to court, the following year.

For now, the head of Longines decides it best that everyone present keep the issue quiet, lest the industry as a whole be tarnished by the scandalous behavior of a single rogue.

"Initial reports home from Antwerp made no mention of the bogus watches. Indeed, the Armand Schwob & Frère display was praised in the press! Even Adrien Philippe's own account of the exposition, published in the Journal Suisse d'Horlogerie, made no mention of the controversy."[7]

Ultimately, a court will find that Armand had sold at least 678 watches bearing the fraudulent "Pateck" inscription before being caught at the Exposition. Armand was ordered to pay 15,000 francs, plus legal costs, to the true Patek firm, and to advertise the judgment against him in two foreign newspapers of Patek's choosing (one ran in The Times of London).[8]

Mind you, those ads were published nearly six years after the counterfeit/fraudulence was discovered and likely a decade or more after it began.

Why would the *Société Intercantonale des Industries de Jura*, the trade group representing honest watchmakers from the mountain region Armand Schwob claimed his territory, not have agitated more strongly and immediately for justice? Why the delayed remonstration?

It couldn't possibly have been because the Swiss watchmakers themselves had something to conceal – that they weren't the earnest, ingenuous artisans they seemed to be?

* * *

Schwob's Antwerp showing – the way it wasn't quite shut down by a jury whose chairman was the victim of the forgery himself – reveals something beyond one man's audacity. The trade would do what it must to protect itself from disruption. Candor, ethics – subordinated, when necessary, to the business.

And so in 1876, when some of these same men heard that America might take their horological lead, they did not convene a committee or publish a white paper. They sent a spy.

* * *

Philadelphia

1876

Theo Gribi delivers a lecture at the Delaware Institute of Mechanic Arts, in Wilmington, on the pendulum – notably how its weight does not affect its rate of oscillation but rather the length from which it's hung, owing to the laws of gravity. The audience marvels at this – and the related fact that two objects of differing weights fall to the ground at the same velocity (again, owing to gravity).

A Swiss watch merchant trying to make a buck in the new world, Gribi originally hails from Neuchâtel – the same area as a later crook by the name of Armand Schwob – but having come to America at 18 in order to sell watches made by his native land's Borel and Courvoisier brands,[9] he meets with ill health[10] and must return to the mountainous clime, that salubrious air, of his homeland.

And yet somehow, the purity of his environs having lent him just that vigor his frail body required, he finds himself able to return to the United States in its centennial year – and to the city where the nation was born – Philadelphia, home to a grand celebratory global exhibition.

And not only that, Theo Gribi has found such vim that he's now leader of the entire Swiss delegation to the horological element of the fair – the larger star of which is a 45-foot-tall Corliss steam engine whose 1,400 horsepower electrify all the other exhibits in its hall – across 13 acres, via a mile of ducts.[11]

That engine powers several machines belonging to the American Watch Company of Waltham, Massachusetts – the crown jewel not only of the US timekeeping industry in this moment but of the entire world.

Walk the floor of the Waltham exhibit and the scale of the thing announces itself. At its center stands Charles Vander Woerd, the firm's mechanical superintendent – a Dutch engineer so brilliant he will be said in his obituary "to have invented more machinery for the rapid manufacture of watch appliances than any other person in the United States." Vander Woerd's market cap alone doesn't convey the man; his hands do. He literally devised and created the automatic lathe in 1871 – two years before it was ever patented – the machine being such a revolutionary tool for watchmaking (and general machining) – and corporate boon for Waltham – that he and his bosses decided to keep it a secret even from the patent office. "Even within the company, very few were permitted to see it."

No matter that particular furtiveness, Vander Woerd ultimately accumulates 27 patents in the field of horology in his day, involving springs and regulators and stems and a wealth of other minuscule components.[12]

But the lathe is something special, in that it allows for the automatic production of tiny watch screws ingeniously – tooling the sides of a spinning rod to create the threads (grooves) of the screw, as well as its thicker head, and then unleashing a tiny circular saw upon that head, to incise the screw's perfect, bisecting slot.

No firm in the world has a single machine capable of all that. Vander Woerd's automatic marvel churns out 8,000-10,000 screws daily. A single worker can oversee six machines at once.

(Whereas a mechanical lathe, such as the Swiss use, remains hand-powered by a single full worker while nevertheless also requiring the assistance of a boy with tiny hands and stands capable of generating but 1,500 new screws daily.)

Beneath Vander Woerd: 800 employees – 400 men, 400 women. They turn out cases almost exclusively in silver (stakeholders in the company having kept the manufacture of gold cases to a separate New York plant) – 50,000 silver cases per year, 100,000 movements – the mechanical engine of the watch, the series of keys, springs, gears and levers whose Rube Goldberg cascade compels each moment's simple tick. That output amounts to perhaps $1.5 million in sales on the open market (prices fluctuating with those of the silver case). The firm's market capitalization: another $1.5 million – staggering for a watchmaker.

Two years before the Centennial Exposition in Philadelphia, Waltham establishes a subsidiary in London. One year before the fair, the Boston Globe, in what may seem jingoistic at the time, writes, "The truth is beginning to dawn upon even the Swiss, that the American system is vastly superior to the ancient system upon which they still work."

Ten years later, a former editor of the New York Times, gazing upon the automated factory, will write, "I found myself before one of the more remarkable industrial establishments ever built; the largest watch factory in the world, the oldest in the United States, the most perfect in every way, yet devised by human genius."[13]

Now at the expo itself, powered by that 45-foot-tall steam engine, Waltham's amazing screw-producing machinery and general horological brilliance delight tourists and astound two foreign travelers, one of whom, as we've already said, despite past frailties, leads the Swiss horological industry contingent: Theo Gribi.

The numbers tell the story as plainly as the machines do: in the four years preceding the expo, Americans have gone from importing 366,000 Swiss watches per year to a measly 75,000.[14] The Americans recognize the quality of their domestic product. What's more, so do the Swiss, says the only Swiss member of the 11-person competition Philly watch jury:

"I personally have doubted the competition," he says. "But now I have seen – I have felt it – and am terrified by the danger to which our industry is exposed."

This same juryperson, Edouard Favre-Perret, himself a watchmaker, will write additionally, "It is time for the old world to hold its head up high to the new, and very soon, so that we are not obliged to say, 'It is too late.'"[15]

Yet, in many ways, Swiss sales declining, its technology utterly lacking, even the exponents of change not understanding the difference between general automation and higher-level machinery capable of performing multiple tasks, it is indeed getting late early.

Waltham walks away from Philadelphia with four medals. Two years later, a Gold Medal at the Paris Exhibition. A year after that, six prizes at the Sydney Exhibition. And then an ad in The Sydney Morning Herald – literally at the other end of the world from the factory itself in Massachusetts – that underlined the point:

"Higher Awards than any Exhibitors in Horology ever received."[16]

A reputation well-earned and to the antipodes spread. Given the immense technological advantage possessed by one side of this industrial competition, it seems almost impossible that the other can do anything to catch up.[17]

So long as everyone plays by the same rules.

* * *

St. Imier, Switzerland

1875

Just 16.5 miles north of Neuchâtel, putative base of the Patek counterfeiter in just a decade, lay the verdant meadows below the town of Saint-Imier – site of Ernest Francillon's watch plant, a*long*side the River Suze – which will soon take the name Longines, after the very long pastures in which it's nestled. This is a plant disconnected from modernity and yet at the cutting edge for the region – which is to say, it's mostly untouched by the true power of the industrial revolution: It has no linkage to any electrical grid nor is it serviced by railways that might deliver coal it could burn, and its hydropower is rather paltry: it sluices the River Suze through a 10-horsepower water wheel to fuel everything.

Ernest Francillon gathers his compatriot horologists to unveil his much-renovated concern, of which he is surpassingly proud – and for good reason, he believes (10 horsepower or not).

Sure, it's all verdant meadow and lazy, languid waterway. But Francillon sees much more:

He tells his colleagues: I have gathered craftsmen from all over the rugged Jura region, each of whom formerly assembled each part of the watch himself, by hand, alone, in his own domicile (for they all used to be farmers who made watches as a side-hustle alone); and, congregating them in my converted facility, I've tasked the group with *collaborating* on chronometry, such that no one man is *ever* responsible for the output of an entire timepiece but can focus instead on a constituent part. So that they can build a better, more consistently well-tooled and finished offering – and with far greater efficiency than their solitary labors allowed. This is a *Société* vertically-integrated, the human hive.

And to facilitate my new collectivism, adds Francillon, I've for nine years tasked Monsieur Jacques David, head of technical development, with the creation of machinery that can assist in the handiwork.

As Francillon's tour of Longines comes to an end, however, his fellow magnates are hardly impressed, least by the machines of Jacques David: Their stated conclusion, which they will later put into writing, even if they don't speak it so bluntly to Francillon's face now:

"At Longines there are none of the complicated machines that are used in American factories."[18]

But they take this argument a step further – and to their own soon-realized detriment: "Their real purpose (that is, the machines' at Longines) is rather doubtful, especially in Switzerland, where it is easy enough to find plenty of intelligent workers."

With this confused conclusion – that individual intelligence will always trump machinery in the age of industrialization – the *Société Intercantonale des Industries de Jura* seals a fate it will soon rectify by subterfuge, espionage and *sheer cheating* – thus seizing the prestige watch market for the following 150 years.

In 1876, at the exhibition in Philadelphia, delegate Theo Gribi, sent by his Swiss superiors, comes face to face with the great ingenuity and capacity of the Waltham watch assembly line – one which timepiece enthusiast Henry Ford will later observe back in Massachusetts – and which will inspire the father of the Model T to adapt Waltham's automated assemblage for his construction of cars (according to Ford's grandson himself).[19]

What can Gribi do in the face of this crisis but scribble a missive back to the trade group in Switzerland: "It must be recognised that we have been left behind by our New World competitors," he writes, "and any Swiss manufacturer who comes here for more information will, without prejudice, be convinced of this point."[20]

The letter lands in the Jura mountains like an alarm bell in a valley of sleeping watchmakers. The Société has already made one error – touring a Longines plant whose failings it couldn't even begin to grasp. For one, Waltham's greatest strength is the interchangeability of its parts, so that not every element must be unique to the piece it ends up in, but parts can be swapped between models. From Gribi's dispatch it understands it is beholding a behemoth in the New World that will put them all out of business. The conclusion is swift: an industrial spy must be sent, lest their business perish.

They choose Longines' technical developer Jacques David – despite his failings as a machine-maker to this point – to report on Waltham, to capture by any and all methods the wizardry of the US watchmaking operation. That the depth and breadth of this scheme will not be fully revealed until the cumulative work of Richard Watson and Aaron Stark is composed and released 125 to 150 years later, the former having sourced and translated secret French-language reports long-hidden in the Longines archives, is a testament to the slyness of the Swiss enterprise: Masters not only of papering over anomalies but of insisting the aberration was in outside perception; of course, the Swiss are the globally-recognized masters of luxury watchmaking – it has always been so – except it hasn't.

* * *

Philadelphia

1876

On their mission in America, Jacques David and Theo Gribi recruit a former assistant superintendent at Waltham – recently retired, he's attending the centennial exhibition in Philadelphia as a tourist – to be their inside source of proprietary company information – technical measurements, automation hacks, financial info – the lot.[21] His name is Ambrose Webster.

Jacques David writes back to Ernest Francillon of Longines: I've tried to figure out Waltham's finances, and "Mr. W, the former director on the mechanical side [is] helping us in this respect."[22]

But recruiting a former employee barely scratches the surface of the espionage that ensues on and off the grounds of a summer festival. For one, there seems to be a literal quid-pro-quo. Not long after Ambrose Webster begins supplying info on Waltham to Jacques David, the former superintendent of Waltham unretires and sets to work for a small watch machinery concern of just six employees. The company soon hires a further 80 bodies. David recommends the firm to his Swiss bosses in a letter, calling the company the best watch tool producer in the US (it wasn't).[23]

Do we have direct evidence of a contractual arrangement between the men? Certainly not. But the brilliant historian Aaron Stark finds "probably evidence of a fulfilled arrangement." And per Ambrose's own biographer, by 1892, he had visited every Swiss watch production plant.[24] Hmm, all this from a man who showed up in Philadelphia a retired tourist?

(I put all this in an email to Longines' official brand historian and archivist Daniel Hug, who, in failing to refute any single aspect of any of it, did advance his opinion on whether the actions mentioned in this chapter constitute industrial espionage: "I just read your lines about Longines and Jacques David, and I think you present them in [a] rather speculative way, which is for many reasons questionable.")

* * *

Waltham

September 1876

Jacques David manages to sneak away from the expo and his consultations with Gribi and perhaps Ambrose, too. He needs to see the plant in action for all the numbers and designs to attain in his mind the vivid outline of what his Swiss overseers must implement to survive. David admits to disguising himself, writing later he entered the plant "incognito."

"Probably entering as a typical worker, he had to ensure he was inside before the company locked the gates at the beginning of the workday, which occurred precisely at 7 AM."[25]

David notes well the way the factory accounts for the time and cost of everything – every move the labor makes and each widget each worker handles ("Waltham operated using an antecedent to scientific management. Foremen time-costed precisely every input that went into each part so the management could know exactly the cost of production down to each piece"[26]).

David is amazed at the way the foremen take note of each workman's momentary absence from his bench – more for record-keeping and measurement than any sort of cruel disciplinarian streak. There is also an amazing lack of smoking and drinking on the job – none! – and a general genteel spirit among the laborers that belies the nature of labor itself (at least as a European who has read Marx might imagine it). "The greatest courtesy reigns in the relationships between all personnel," David writes to his Swiss dispatchers. "The observation of these simple details facilitates the work of everyone."[27]

Even the décor struck him as conducive to a spirited workmanship the Swiss could only aspire to: "The most rigorous cleanliness is maintained, and this care is encouraged by the pretty pieces of furniture, the offices, the lighting and the very appearance of the benches which furnish the rooms. Some workshops are decorated with curtains, paintings, photographs of the workers, flowers, etc., which make being there pleasant."[28]

Of course, even the most soothing ambiance could not a superior tech gadget make – consider all those early-aughts stories of the office amenities of Google and Facebook – no matter their number of ping-pong tables or bean bag chairs, the engineers still needed to compose innovative code, or else the whole of the venture would be sunk and empty (think WeWork).

And so Jacques David, during his run through the factory in disguise, manages somehow, perhaps with an accented charm, a continental charisma, to turn at least one, if not more, Waltham employees against the firm that has ostensibly treated them so well.

"We have inside sources and we shall soon have the information we want," he concludes to his Swiss compatriots.[29] He is determined, in particular, to ascertain how Waltham manages its affairs so efficiently and without financial stress – what are its costs of production – its *cost per unit*?

***

Société Intercantonale des Industries de Jura

1877

Jacques David's two final reports land on the desks of the great Swiss watch proprietors – addressing Ernest Francillon, head of Longines, perhaps most especially. If anyone should survive and succeed, it's his good friend from St. Imier. The reports include, for instance, the number of watches Waltham produced in October of 1876 (6,500) and their cumulative value on the market ($63,000) – the kind of proprietary intelligence no competitor should possess.

The Société handles the documents like contraband. Eight facsimile copies are distributed to its component members; the originals disappear into a vault. The spoken rationale: should Waltham ever see the investigation, it might just use it in advertising[30] – *See, even our Swiss competitors believe us better.* It is not untrue that advertisements are perversely boastful in this era of industrial competition. Naturally, the Swiss horologists have a far better reason for keeping proof of their unethical, if not totally illegal, surveillance under wraps – and indeed, all paperwork indicative of the spying won't be seen again until the Longines archive releases a copy in 1992 (that year, Longines celebrated its 160th anniversary, and while the release is to its credit, I'm not sure it had any idea quite what it was admitting).

Francillon, tremendously indebted to David for his American deception, ensures that his spy becomes director of Longines in 1880 – just four short years after his Waltham factory breach.

"Jacques David, Théodore Gribi and the others were the principal agents for the modernization of the Swiss watch industry," historian and Jura-region archivist Jean-Marc Barrelet will conclude.[31][32]

* * *

Waltham

1877-1957

When you lose a marked advantage in intellectual property – powered in the first place by shrewd businessmen and even cleverer inventors – your room for error narrows – like walls closing in on a filmic hero, one with only seconds to figure an escape – and the firm's future success depends on those c-suite deciders and mechanical thinkers who inherit a venture now squeezed.

Waltham can be said to have made strategic errors – focusing on lower-priced watches at a time when it could have still stolen the luxury market completely from the Swiss, for instance; but in reality, it wasn't the consequence of any single choice that it was left behind — but of repeated poor directives, the decisions in aggregate of a few uninspired , if not utterly unworthy, deciders.

There was Ezra Fitch (not the same Ezra Fitch who partnered with Abercrombie, though it's an uncanny coincidence), who as superintendent from 1884-1923, president from 1886 and board member all along wielded too much power – especially since his philosophy was to assess the market and then cut costs so as to give the masses what they wanted cheaply (could've worked; didn't; "Fitch did not inaugurate the policy of catering to the public fancy, but he gave it much wider scope").[33]

A Saturday afternoon, November 1884: Waltham management gathered the plant and announced that by December 1st, a large chunk of the 2,600 workers – probably 500 – were to be let go, the rest to have their salaries slashed 10 to 20 percent. The news landed like a gate slamming shut. May 1894: the turncoat Ambrose Webster, the retiree who'd spilled his secrets to the Swiss at the fair, died on a work trip to France – still as unretired as he'd been since he met Gribi and David. 1924: wage cuts of up to 40 percent triggered a nasty, protracted strike that crushed production for five full months – after which some of the best employees who'd walked the line declined to return and took jobs elsewhere.

Then the floor gave way. Bankrupt in 1948. Bankrupt again in 1950. A government loan of $4 million dragged the company back to its feet in 1952 – only for it to wind up, four years later, in the hands of an early corporate raider, a firm called Bellanca, about which Time Magazine ran the headline: "How to Loot a Company."

Lucky for Waltham, Bellanca gave up on sucking Waltham dry speedily. On the same October day Don Larsen threw a perfect game in the World Series, vaulting the Yanks over the Dodgers and catcher Yogi Berra into Larsen's arms in an embrace whose photograph occupied four columns on the front page of the *Boston* Globe, Bellanca, via a bank intermediary sold Waltham to a syndicate led by Diamond District Joe Axler – a Baume & Mercier importer who owned a share of a ritzy Miami Beach hotel visited by the Shah of Iran.[34][35]

Axler – the sixth management change in ten years – promised to revive Waltham's legacy consumer business. He promptly spun off the watch division from the other timekeeping enterprises (by this point, 90 percent of Waltham's income came from gyroscopes and aviation equipment). That former enterprise to which he claimed so much devotion? Two weeks after the spin-off, he licensed the right to produce Waltham watches to Hallmark.

In the late 1960s, the head of Hallmark "accepted an offer from Swiss suppliers seeking an American outlet for their watch mechanisms" – or, in other words, he surrendered the historic American firm finally and fully to the Swiss society brands that had infiltrated it so malignly nearly a century earlier.[36]

(It has since changed hands again and again – so that my grandfather's gift watch came from a very specific place and nowhere at all.)

***

New York

2025

I purchase two watches as I complete what I consider an essay about the taking of something that can't really be possessed.

And yet everyone, even those unaffiliated with high horology or lowdown dirty tricks, knows what it is to steal time: Its remove from another's existence a quantum that simply cannot be recovered. There is no medicine, no Blue Zone protocol for long-life, that will return to us a moment that has passed.

As I write this, the world's oldest person, a Brazilian nun named Inah Canabarro Lucas, age 116, has just passed the prior Wednesday, five days earlier. The good sister was apparently an enthusiastic fan of her local football club – Internacional, of Porto Alegre – whose jerseys she donned, per her Times obit, "on milestone birthdays."[37]

Born before the sinking of the Titanic, she died in the age of AI – and yet her obsession was a game bound by just 90 minutes, in which players are charged with slotting in a goal before a final whistle. Instants and opportunities can be taken away – scoring chances, one might say in sport; but that they are finite and not unlimited, that they cannot last, is what imparts their electricity. What brought Sister Inah, across her long stay on earth, to life.

And if by some odd physics, weird things we can only claw our way to understanding now, these Polaroid shots of life, precious moments whose lone remnant in the memory is what makes Alzheimer's so utterly devastating to consider, are skewed one day, so that time seems yet more bountiful but also more elusive, subject to compression and dilation both (see "Interstellar") – it seems impossible there won't remain some form of its taking – quanta whose losses we will feel as if time is tactile, whose misappropriation we will lament as we drift back and reminisce.

And for that reason – the symbolic restoration of stolen time – the recapture of occasions industrial and personal both, I purchased two watches, as I said, in writing this. One, on eBay, the same fake-gold quartz Waltham gifted to me by my grandfather in a scene I can remember but never relive (or a model that looks nearly identical). The man in the Midwest from whom

I purchased it refused to mail it, however – for reasons unknown to me but in a fitting twist – and so eBay refunded me fully (a bit of fairness unusual in this tale).

The other, because I'm perhaps a vengeful person at my core, though I recognize it's a terrible quality, and I like pursuing forms of payback even on behalf of others – plus I feel a coreligionist affiliation with Armand Schwob, though he was a reprobate and a fraud: I can't help it.

It's a fake Patek Philippe.

(Schwob's fake Pateks, engraved "Pateck et Cie." now retail for thousands of dollars in auctions and online; as far back as 1992, long before any recent inflation, a Boston Globe ad read, "MENS 14kt gold pocket watch...signed Pateck & Cie. Geneve $10K orig bx"; [38]Schwob himself, before his 1930 death at age 73, had run through and spent a stolen fortune of $30 million and had escaped countless tight spots – including a stint as a prisoner in Sing Sing).

MENS 14kt gold pocket watch (ca. 1830), signed Pateck & Cie. Geneve $10K orig bx 617-734-8424

And the clincher: Armand Schwob's original Paris business testifies to this day to the greater IP theft the Swiss pulled off all around him.

For when Schwob was located in the Jura region of Switzerland, on Rue Leopold-Robert in the late 19th century, he and his brother shared an office building with another pair of Jewish brothers of both the same surname and profession – in a real "Twilight Zone" twist – another two Schwobs who sold watches wholesale under the name "Schwob Freres et Cie." (The coincidence is covered in the March 2018 issue of *Antiquarian Horology).*

Among the timepieces these doppelganger Schwobs vended? Complex chronoscopes (essentially, a specific type of chronograph, one usually equipped on a single dial with a tachymeter, a pulsometer, and a telemeter besides simply a timekeeping function) and repeater watches (those that chimed at various intervals so the timekeeping is made audible).

These were produced most proximately by a man named Prosper Nordmann – a local who'd lived in the States and worked in the horological sphere there – only to return to Geneva and continue plying the same trade under the very functional corporate brand "Timing and Repeating Watch Company."

The watches he produced after return*ing* from America – those that passed through the hands of the doppelganger Schwobs in *the same office* as Armand before reaching retailers and then consumers?

Clones of watches produced by the American Watch Company – of the wares carefully constructed in Waltham, Massachusetts.

ARMAND SCHWOB & FRÈRE

14, Rue Léopold-Robert, 14

Maison à Paris : 19, BOULEVARD BONNE-NOUVELLE, 19

Horlogerie en tous genres. — Achat et Vente

EXPORTATION POUR TOUS PAYS

An ad for Schwob's business when it was located on Rue Leopold-Robert above and the actual office shared with the other Schwob pairing, who retailed cloned Walthams, below:

The fake Waltham watches – copies of chronometers from the factory spied by Longines – reached Schwob, the man who'd tried to bamboozle Patek Philippe – it's the perfect illustration of the Pandora's box Switzerland opened once it decided another country's industrial method could be quietly lifted, translated and turned into its own advantage. History bears out this karmic punishment for time-taking:

We know this, in part, by the work of a single thief – a man whose kleptocratic obsessions led him to steal the most valuable watch in the world – the one originally designed for that ill-fated aristocrat whose monarchical despoiling of her people we summarize with a single quote: "Let them eat cake."

Taking begets taking.

Tick, tick, tick.

2

The Origin of Everything: France, 1800

If the question is, "How have the Swiss managed to dominate watchmaking across centuries of turbulence in every other industry?" then we must take a step back and ask: Where in the first place did they acquire that expertise — the starting point from which they emerged to automate production, to modernize, to exploit rival firms' planning in the 19th century?

Who provided the plans that became the foundation of their great national project?

Enter several generations of ingenious English inventors beginning in the early 1700s. Their great challenge back then is really the same one watch designers grapple with today: figuring out ways of releasing at even intervals tensed, poised, coiled-up power — which would naturally prefer to explode into action in one spasmodic, tantrum-like unfurling, energy always seeking its lowest state — via an "escapement" and with the least amount of friction (and therefore, the least undesired loss to the ether of that initial power source).

Many forms of such regulated release were tried and tested. George Graham, who lived from 1673-1751, toiled over his deadbeat and cylinder escapements. His student Thomas Mudge outdid him, inventing in 1755, just four years after the master's passing, an escapement more clever and longer-lasting than anything his teacher had ever devised (we still use it in nearly all automatic wristwatches today): the lever escapement — a forked pallet whose bejeweled tines intentionally, while oscillating, become stuck-and-unstuck (with regularity) in the teeth of a rotating wheel with table-saw-like serrations — curled outcroppings like jacket hooks on the back of a door or Hokusai's wave (which the Japanese woodblock master would not even create for another 76 years).

Meanwhile, John Harrison, solved the longitude problem — how to create a marine chronometer that would maintain accurate time at sea (where oscillations are goofed up by the ocean's own motions) such that a ship's longitudinal position could be calculated (this quest is the subject of an entire book by Dava Sobel). Arnold and Earnshaw then worked to refine and commercialize the chronometer.

But the century-plus of English innovation came to an end just as a Frenchman with a superior mind for timekeeping appeared on the scene. The Mozart or Messi of timekeeping. The innate genius no Salieri could ever hope to match (per that fictional film's rivalry, if not history).

And it was the great thinker's conceptions, contrivances and creations that provided the true base of knowledge upon which the Swiss built (while eventually failing to share with you, the general public, the prime importance of their prime predecessor tinkerer altogether).

Yet this is no mere chronological rewind — the tale of the man whose lifework gave Swiss horologists their intellectual start, the knowledge base they needed before they ever could surveil Waltham. It's almost eerily a prequel story of IP thievery.

But what was stolen was not a great group's or single rogue's novelties in marking time – unique insights in how to measure it best– it was *the object* in the world that most comprehensively realized those concepts. Brought them to material life. A mechanism possessed of one man's best ideas, its entire existence a byproduct of his mind.

This is the tale of Breguet — the modern progenitor of all Swiss horology though he was not Swiss, who put all those seminal-to-the-Swiss ideas into a single timepiece for Marie Antoinette. Which was centuries later snatched.

* * *

Abraham-Louis Breguet's timepieces are not during his lifetime assembled in Switzerland, however badly the Swiss conglomerate presently controlling his name as a marque might wish it otherwise.

The story of the (Prussian-French) man and the company that took his name is one of invention and exile, exhibition and disappearance (over and over again).

A Breguet watch leaves its mark, in part, when it's taken, by the great absence of what once was. In the void. This is also how the man himself came to his expertise – via sudden loss.

Abraham-Louis Breguet's father, an innkeeper, dies at the age of 39. Not so unusual for his time – if a French male born in the middle of the 1700s in France survives till 20, he can expect to live till 50. But Breguet is but a lad of 10 or 11 years old in 1758. So his mother remarries hurriedly a cousin of her deceased partner – and in the particular match the entirety of watch history as we know it still today, falls into place. For Suzanne-Marguerite weds Joseph Tattet – a watchmaker, who promptly dispatches Abraham-Louis to Versailles for a horological apprenticeship.

The death of an innkeeper, the loss of a father, yields the union of singular intellect and the craft perhaps best suited to that imagination and resourcefulness.

Breguet is born in a town that's now Swiss but is then Prussian. He's destined to spend the vast majority of his life in Paris before dying in 1823. That is, he will live through the *Ancien Regime*, its overthrow during the French Revolution, the Reign of Terror of the ensuing guillotine-guided ad-hoc-government, the empire of Napoleon and then finally that man's Waterloo and the monarchical Bourbon Restoration.

Breguet the man has a deeply receding hairline that leaves a pale, prominent forehead, doughy like a newborn's tummy. But his faculties as he sets to work professionally are not in question – evident to even the most discerning and patrician crowd.

Stendhal writes in "Rome, Naples, and Florence":

> *The moral sciences show so much evil in this creature known as man—or rather (which comes to the same thing) it is so pleasant and so tempting to think of him as being better than he is*—that the imagination can but rarely find a taste of happiness unless it take flight into a world remote from reality. Breguet can devise a watch which will keep time unerringly for twenty years; whereas the gimcrack mechanism through which we live breaks down and gives pain seldom less than once a week.[39]

To be clear, Stendhal is referring to Breguet's eponymous firm, but that company merely continued to carry out the man's ideas and designs and instructions.

Breguet is a focused soul – a man whose work could have easily yielded him all the gaieties and indulgences, all the pleasures and perks, of French high society. A man who seems to have forgone that silken, champagne-soaked life for the drab lab coat and highly-functional work bench of a technician lost in his thoughts. A man devoted to tinkering with microscopic parts until those parts seem to loom very, very large. You know those adults who fall so deep into a hobby that it swallows up their souls? A chef, a guitar-obsessive, a garage-owner gripped by the engines he builds and rebuilds?

Someone always putting together puzzle pieces in his mind.

Yet Breguet somehow also has social skills – the sort of temperament and savvy required for a global business rife with schemers and bottom-line-focused businessmen. Breguet manages to get his watches distributed across the world. But when a business partner named Gide, determined to milk Breguet's genius for every penny he can in the marketplace, begins to harass him via letter – to abuse him verbally, as Breguet was in the hobby of making one-off watches instead of iterating successful past designs – Breguet does not take the bait or reply in kind. He merely buys Gide out of his share.[40]

And perhaps it is a window into Breguet's mind that while his watches featured fluting and guilloché, those elements, employed ever since for ornamentation, were utilized by him for highly rational – *practical* – purposes: the fluting to make the pocket watch easier to grip (ridges reduced to their frictive effect) and guilloché to help the eye distinguish one dial area from another (this is a man whose watches could combine timekeeping with temperature measurement – the eye needed a subtle guide to such information concentration).

What Breguet accomplished – not as a list of patents but as a sequence of problems solved, each one a small defiance of physics:

In 1780, the watch that wound itself – the *Perpetuelle* (long before Hans Wilsdorf of Rolex came up with his own Perpetual mechanism, though Rolex, being a company born partly of ingenious marketing, would prefer you overlook this).

That alone would have been enough for a legacy. But Breguet was not done – not remotely.

The shock absorber for a watch movement – so that the jostling of a man on horseback or the sway of a woman in a carriage would not throw off the timekeeping. The gong spring for chiming watches – watches that enabled you to tell time in the dark merely from sound, where models made by others used bells. The touch watch, which enabled you to feel the edges and ridges of the watch in your pocket and thus tell the time from its protrusions without ever having to remove it, much less gaze at its face – time, literally, at your fingertips.

And then further still. The sympathique clock – a docking station for watches that then transferred to them, entirely mechanically, the clock's own correct time – the sort of synchronization of disparate devices Silicon Valley has chased down for the last few decades. The tourbillon mechanism, whose movements counteracted those of gravity so that the constant downward pull of our environment would not affect the watch's machinery as it does our own corporal mass – in this way, the Breguet watch escaped from Earthly pull 170 years before Aldrin and Armstrong on the moon.

And in 1810, as if to prove that even the vessel containing all this genius could be reimagined, Breguet created for Napoleon's youngest sister, the Queen of Naples, the first watch designed to be worn on the wrist – instead of stuffed into either a man's vest pockets or the special folds of a woman's dress included in that era for the carriage of a wearable clock.

And all of the above brainstorming – executed and commercialized by Breguet himself and thus far more impressive, one might argue, than those visions Steve Jobs had, the execution of which he always left to technically-savvy others – overlooks those componentry upgrades that the average reader might not yet care enough to investigate: Breguet-style watch-hands, the double second watch that presaged the chronograph/stop watch, the balance-spring accuracy-improving Breguet overcoil, the guilloché technique on his silver and gold cases.

All of those revolutionary improvements I have undercut as wonky were truly groundbreaking and soon adopted by others, but they just seem almost secondary when you consider the list above them – that Breguet pioneered the wristwatch (others had tried the format earlier but without real success), the clock that could set your personal watch's time, the watches that could transmit to the wearer time via gong or tactile changes along their edges.

Oh, and before Alexander Graham Bell connected us all telephonically, Breguet made mechanisms for his collaborator Chappe's "optical telegraph" – essentially, a series of towers that could transmit messages one to another over long distances via symbols on one end and a tremendous telescope on the other.

Breguet envisioned a world of hyperconnectivity and synchronization with none of the dystopian cynicism those words now arouse in us – because he himself was the revolutionary – the man who forced the point, so that people in the dark, scattered and unable to understand one another, first begin to interpret the world by shared, advanced device. And was this not a noble goal of the Enlightenment more generally being fulfilled by an individual on humanity's behalf?

Which is why a museum trove of Breguet's treasured timepieces, including one made for Marie Antoinette with every horological complication Breguet ever devised, are the objects targeted – and successfully taken – centuries later in what's universally recognized as the greatest watch heist ever attempted...*and accomplished.*

* * *

The heists come later. But the appropriation of the brand itself – that begins with a bankruptcy filing and a Swiss businessman's eye for a bargain.

A reader might reasonably ask: isn't there a categorical difference between physical masterpieces of timekeeping being taken and watchmaking methodology being illicitly copied by corporate spies? Does the attempted heist of the most complicated mechanical watch ever created (and dozens more smaller plots, of which a few are mentioned) really cohere narratively with the Swiss attempt to infiltrate the American Watch Company of Waltham and convince the world what it stole it had originated?

Well, on the surface, no – stolen ideas and objects – even if both obtained for venal reasons – don't quite seem to belong to the same category of thievery.

The catch is, all the hullabaloo surrounding Breguet's inventions and the great heists of those objects mentioned here is written to underline the man's place as the essential forefather of modern horology.

And while he was born in Neuchâtel when that town was under Prussian control, all of Breguet's horological issue occurred from his Parisian atelier (as a naturalized French citizen). The man left Neuchâtel at 15. That is, the father of watchmaking was not Swiss but French, really – and this poses a marketing problem for the Swiss in subsequent centuries, the extent of which is only made clearer and clearer as we delve into the great Breguet heists – these hugely romantic thefts that play out across decades.

If you are positioning your industry as the originator of high horology, it does not help to have a Frenchman's fame or the infamy of those taking his pieces everyday reported again in the papers.

And there is every indication that the Swiss recognize this marketing challenge about 100 years after they first notice the technological threat posed by the American Watchmaking Company in Waltham.

For instance, the Swiss high horology retailer Beyer, whose retail shop operates in Zurich, on its web page dedicated to Breguet, displays this most contortionist/mendacious paragraph:

"The name Breguet is part of Switzerland's cultural legacy. Abraham-Louis Breguet, from the canton of Neuchâtel, is considered one of the greatest watchmakers of all time. He invented the tourbillon in 1795. He improved the automatic movement in 1780 and since then has been called the father of modern-day watchmaking. Breguet is now one of the few manufacturers to produce all of the parts of its watches itself."[41]

Suddenly, Breguet is Swiss! Horology descends from that country's peaks and valleys, after all.

This is hardly the only marketing that conveniently relocates Breguet's genius within Swiss confines – it has been said for decades now within the industry that the Swiss sell the legend while conveniently forgetting its geography. And this has been replicated in the physical world itself:

Paris, 1976. The quartz crisis is gutting the mechanical watch industry – the piezoelectric oscillation of a slice of quartz stone, battery-powered and hooked up to a stepper motor, has suddenly become a widely available and more accurate timekeeping technology than pent-up, wound springs and mechanical gear trains. Watchmakers are essentially rendered useless overnight. François Bodet and the other executives at Chaumet, the French jeweler that owns Breguet, look at their increasingly empty Parisian workshops and face a grim arithmetic: if the company is to stay alive and committed to mechanical watchmaking, it needs to go where the mechanics remain.

So Chaumet does the unthinkable – moves the company whose founder signed his dials "Breguet à Paris" entirely out of France, resituating it in the Vallée de Joux in Switzerland. Bodet describes the decision in economic terms alone – pure survival, never intended to challenge or alter the long and worthwhile history of Breguet, the brand and man, as purely Gallic.

And yet, to the credit of their business acumen, one supposes, the Swiss sense an opportunity here in this French brand's misery: Nicolas Hayek, in particular, head of the Swatch conglomerate, realizes he can buy on the cheap the name and great history – the archive – of the Breguet brand (and in so doing, the legend of the man himself) and then resell that package how he pleases.

And how does he decide to do it, after Swatch buys Breguet from the insipidly named Investcorp (successor owner of Chaumet – and a company based in Bahrain, of all places)?

Well, when Abraham-Louis Breguet produced watches, he signed them, "Breguet à Paris."

Under Swatch? "Swiss Made" or "Swiss guilloché Main."

That is, the more we explore the great allure of the Breguet timepiece to a crook, the more we understand why the Swiss themselves so badly want power over the brand, which was once a dazzling gem but for much of the 20th century has been fallow – a brand name owned by jewelers named the Browns (when the Bahrainis tried to make Breguet watches in the early 1990s, they first had to install movements from another company they owned – Nouvelle Lemania – movements that could be found in Omegas and Pateks).

But vintage and classic Breguet is beauty, mechanical-genius-made-visible, complexity beyond what humans assume other humans can engineer or achieve.

It is legacy. A kind of world record made manifest in mechanism.

It's our collective ambitions, glinting in a glass case.

Just waiting for us.

* * *

Before the thief — one Naaman Diller — enters the picture, there must arrive the legitimate collector – equally enthused by the lure of the object.

Enter Sir David Lionel Solomons, a Baronet by way of title inheritance from his uncle, a lawyer by virtue of study and the conference of degrees but a madman tinkerer by dint of his unchanging nature.

Solomons is the nearest thing rural Kent has to an Edison: Solomons has a massive workshop in his home powered by a coal-fired generator capable of producing "enough electricity for 1,000 sixteen candle-power (about 60-watt) light bulbs."[42]

And Solomons is fascinated by the lightbulb – his is one of the first houses to be electrically lit in all of England. And by magnetism – he possesses an electromagnet of such strength it is often borrowed by his scientist friends, of which he has many, including neighbor William Siemens (yes, that Siemens, the German-British inventor whose surname lives on as a brand) as well as Lord Kelvin (he of Absolute Zero).

Solomons is also greatly engaged by the workings of the newly created automobile – he owns the second such car ever registered in his country. And by photography – his estate has a darkroom and a studio and he projects his work in a theater on the estate grounds for the public – including some moving pictures via an early mechanism called the kinematograph.[43]

"His home in Tunbridge Wells is a modern magician's cave," writes Vanity Fair in 1908, "with electricity to play the part of the good genie. All sorts of weird things happen there at the touch of a finger."[44]

He also researches the lead-acid batteries in direct current powerplants, which both store any generated power as accumulators but also adjust the voltage ultimately emitted, to such a degree that he authors a monograph on the subject.

So when Solomons begins assembling his collection of Breguets it is with the fascination of a fellow engineer, of a man whose own book examines the aggregation, storage and dispersal of energy – which are the three essential actions undertaken by any mechanical watch, in point of fact.

Solomons assembles a collection that includes works preceding Abraham-Louis Breguet's but features most prominently 57 watches of his famous manufacture, including his masterpiece, the *Perpetuelle* Marie Antoinette, which the brand labels numerically as model 160 – a watch commissioned either by one of the queen's guards or a Swedish nobleman named Count Axel von Fersen (depending on which story you choose to embrace) in 1783 and not finished until long after she has been killed and Breguet himself has died – by Abraham-Louis' team, in 1827 (four years after Abraham-Louis' passing).

The order for the watch requires it be made entirely in gold (except where steel must be used) and contain every complication (that is, added feature, novelty or module) then known to horology.

And so it does: within a 63 mm-diameter case, the watch (whose workings are visible – making the piece "skeletonized"—thanks to a crystal case highly unusual for the era) features the equation of time (the difference between mechanically-kept time and solar time, the latter of which is of course subject to shortening and lengthening throughout the seasons, as our planetary home spins and rotates, such that high noon is not always at noon per measurement); a power reserve indicator (also known as *state of winding* – a gauge of how much winding the watch has yet preserved and therefore how much power it still possesses before it exhausts itself; or, looking at it glass half-empty, how much winding it still requires to be fully charged – thusly primed this one can run without stop for 48 hours); a sector displaying the date of the month; a white enamel main dial in Roman numerals for the hours and minutes and a smaller sub-dial (short for subsidiary dial) that features along its circular rim not only markings for seconds but also, at regular intervals, the days of the week – so you can look at the seconds and learn simultaneously that it is Mardi or Lundi; and, finally, a thermometer, whose scale is derived from the different expansion rates under heat for precious metals, which are soldered together and twisted into a helix here).

"It has been described as the most famous watch in the world," the New York Times says in January of 2025, that is, 198 years after its completion, a period that has seen the invention of phones, planes, computers and the internet.

"Some say it is the most valuable."[45]

And, after Solomons bequeaths it to his daughter and she in turn gifts it to the trustees of a museum she pays to create, it is the loot, the bounty, the take in the greatest horological heist ever conducted.

* * *

Walk into any auction house during a major watch lot and you will notice a strange incongruity at the heart of the mechanical watch world – the more distinctive, unique, *singular* a timepiece seems to be the more aficionados, and even the public at large, seem to laud it (the Marie Antoinette *Perpetuelle* is the subject of a 2025 museum exhibit in London, and it is the first time the piece had been loaned by the museum that owns it; in fact, as the piece is considered priceless, it is nearly impossible for the recipient of the loan to insure it, because any valuation would be comical, though an undisclosed number is ultimately agreed upon and fixed for this removal and traveling display).

Yet if you look at Breguet Number 2788, created in 1818, you'll see that singularity in timepiece is a backwards concept – somewhat akin to a single person adopting the internet when only a mass of adopters would make the technology useful. See, 200 years ago, Breguet has a theory – that were you to put the engines of mechanical watches near each other, they might benefit from the company – the inexactitude of one might be mitigated or rectified by the workings of its neighbor. And so he actually goes ahead and installs within a 63-millimeter diameter pocket watch (made for the future King George IV, now just a prince) two entire mechanisms on a single plate, observing subsequently that the two separate shimmying balance wheels set the whole plate "in continuous microscopic motion...and that any errors in one half of the watch would be favorably influenced by the other."[46]

As it turns out, there is greater accuracy in a watch *not* being alone and apart – not being singular – in a law of averages smoothing out mistakes. This is the incongruity. This is also the principle underlying developments centuries subsequent to Breguet's experiment, when companies release in the 20th century (and still do today) so-called high-beat watches, in which the balance wheel may beat 28,800 times per hour or even 36,000 – rather than 21,600 times, which frequency is considered more of a simple baseline (a quartz watch by contrast beats 32,768 times...per second instead of per hour).

But the logic of these increasing numbers is simple – the more oscillations (or two beats – like a pendulum swinging back and then forth), the less any single disrupted oscillation (due to a nudge or shock or some small incident that temporarily messes with a vibration) can alter the watch's accuracy. There exist so many more non-disrupted back-and-forth swings for the single error to stand out.

Watches very much adhere to the math of Harry Nilsson, in counterpoint to our fetishizing (my own included) of the singular masterpiece by the supreme genius: "One is the loneliest number that you'll ever do / Two can be as bad as one / It's the loneliest number since the number one."

There are master watchmakers today who resist or rebel against Nilsson numbers – most notably, Roger Smith, former pupil of George Daniels, the autodidact English watchmaker of last century whose description of the double-mechanism-single-plate pocket watch I earlier quoted (indeed, he was invited to co-write an entire book on the Sir David Lionel Solomons collection of Breguet pocket watches); Smith believes you should ensure your exactitude elsewhere in the mechanism – that a reliance on higher numbers to even out your accuracy troubles introduces ever more wear and tear with each further vibration.

It's a moot point in the case of the stolen Antoinette watch – for in that case, the museum founded by the daughter of Sir David Lionel Solomons – a museum primarily dedicated to Islamic art for two reasons (one is that Solomons' daughter has been trying to promote interfaith understanding in the land of Israel for decades and the other is that she may be engaged a love affair with a professor of Islamic art at Hebrew University)[47] – is ransacked almost across the board in April of 1983. No singular piece is taken alone – the thieves don't go after just the Antoinette.

One-hundred and six items, the very best watches in the collection, pieces that have passed through the hands of an electrical genius a century earlier and a mechanical wizard-creator 100 years before him, are nabbed that night.

How can you even begin to estimate the worth of *all* that is gone? These machines have no peers in all of known human creation. The Associated Press insists on the arbitrary price tag of $5 million – for all 106 pilfered items.

Museum director Gavriel Moriah sniffs at this, insisting on the incalculability of the haul and adding this anecdote: "He was once offered $1 million and a pledge to buy a Phantom jetfighter for the Israel air force for the *loan* of *two watches* for as long as the unidentified collector lived."[48] His successor as artistic director, Rachel Hasson, calls the Antoinette "the Mona Lisa of the clock world."

Conjecture among security forces soon lands on exactly this type of easily-caricatured (perhaps criminal) magnate abroad: Has the raid been ordered by a mafioso or a sheikh? A Sicilian or an Emirati?

Then the case goes cold – and remains so, unresolved and shocking, for more than 25 years.

* * *

The Frick Collection, Fifth Avenue and 70th Street, November 1982. Winthrop Kelly Edey – a high society collector of timepieces whose family money, earned from grandpop's manufacture of oil refineries, allows him to wake in the late afternoons (say, 4 p.m.) and spend his late evenings tinkering with clocks and writing – has done what no one expected. Your own author feels a great affection for this man due to his routine alone (that he manages to schmooze with friends at diners in the interstices of this routine only makes him that much more likeable).

Beginning at the age of 12, Winthrop Kelly Edey collects fine clocks, and because life presents us with facts better than we could ever contrive, he hooks up with the Frick Collection on Fifth Avenue and 70th Street on the Upper East Side in the early 1980s and facilitates a show of clocks in that sanctuary of fine art. Not only is he not denied this exhibition but he is permitted to curate it and write its catalog, all of which is made available to the public from Nov. 2, 1982, to Jan. 31, 1983.

And the arguments against him are loud. "Clocks don't count as art, some people say," the Times review of the show sniffs. "If clocks are allowed to be art, there'll be no end to it. Knives and forks will be art, and umbrellas and socks."[49] (The review is ultimately not an endorsement of this position but neither is it the least bit well-written or informed; the year is 1983 – Marcel Duchamp converted a porcelain urinal into a provocative "readymade" artwork a full 66 years earlier.)

Which is where things get eerie: Included in the exhibition most prominently are Breguets – including one made not for Queen Antoinette but for another royal caught up in the Revolution.

The Duc de Choiseul-Praslin was at one point a prisoner of the entropic and convulsive coup. The clock Breguet presents to him brilliantly promises him his best chance of survival. For before the Revolution, the French keep the same 24-hour method of measuring time we use today. During the Revolution, however, for between three and 12 years depending on where you are, the terrifying French potentates discard with that system and institute a metric-like system of base-10. Days are divided into 10 hours, hours into 100 minutes, minutes into 100 seconds.

Breguet, knowing full well the need for a duke allied to the old world to survive somehow in the new, makes him a desk clock with two mechanisms whose faces appear on opposite sides – one side reflecting the familiar way of telling the time and the other the decimal system (that the decimalization of time never takes off, apart from its roots in those bloody years, is an interesting subject on its own, considering how every other measure across Europe eventually bends to decimalization, even the United Kingdom's strange system of pounds, shillings and pence in 1971).

Breguet understands that the duke (not to mention his own family) straddles a perilous historical divide: For one visitor to the office, you might need to show one side of the clock to stay in good graces – for another, the clock will have to be flipped.

The only true time on Earth is the one that permits you more.

This is the great cleavage between past and present the Revolution forces upon figures of that era and that Breguet knows must be reflected in his work. Not incidentally, that beautiful clock, a standout of Winthrop Kelly Edey's 1983 show, is later bequeathed to the museum by the eccentric collector (is there any other kind?) upon his death, in 1999.

What Winthrop Kelly Edey and the Frick do not know 16 years earlier, back in 1983, however – this is the eerie bit – is that theirs will be the final museum show before Breguet's legacy is itself separated into a before-and-after. That within a couple months, the master clockmaker's greatest pieces are to vanish.

That Winthrop Kelly Edey, nocturnal writer and amateur clock-fiddler, alongside a museum better known for its Vermeers and Upper East Side real estate, would present the last clocks before the Breguet story is divided into two, like a clock with two faces.

* * *

What do we want out of our watch rogues – that they be in the illicit endeavor entirely for the thrill of money ill-gotten – that they should be flippers who treat watches and diamonds alike, as mere commodities of their trade – or that they somehow, while devoted to underworld tactics, live and breathe that same appreciation for the pieces as legitimate collectors, that they devote indiscriminate time and energy to the study of the mechanisms, the makers, the artistry?

And what to do about the watch thief who is somehow both: an obsessive about the details of an irreplaceable piece while also, historically, a compulsive taker of any and everything?

Enter Naaman Diller – kibbutznik, kleptomaniac, mastermind.

Who singlehandedly extracts all 106 pieces from the Jerusalem Islamic Art Museum – including those clocks that require disassembly to fit through metal grates he pries apart with a hydraulic tire jack (and then reassembles once he has escaped).

Enter Naaman Diller – whose very theft betrays his intricate understanding of Breguet's complications.

* * *

Raised on Kibbutz Ein Hahoresh, Diller is a thief from the jump – a kid with an unshakeable compulsion to take from others. His exploits eventually earn him the nickname "the Kibbutznik burglar." He takes from his fellow collectivist growers – stamps, vacation money, even the literal fruits of their labor, which he hides under his bed. *Guavas.*

And yet, he is not simply amoral or without compassion. He possesses from an early age a complexity belied by his childhood penchant for silence. The son of an educator and a psychologist, Naaman Diller – who oversees an animal farm while not studying – is regarded by his kibbutz contemporaries as a strange genius not quite able to form social bonds.

Once, his entire cohort heads to the movies in Tel Aviv – only for Naaman to claim halfway there that he has forgotten an item on the kibbutz, return to that community and burgle the very kids with whom he should be bonding at the cinema. Then he returns to the theater and watches the film. As if somehow the two activities are not utterly incongruous.

Years later, after someone on the kibbutz dies, Naaman departs the deceased's funeral and burgles the homes of his neighbors then grieving at the service. When he is later caught – and Naaman Diller never seems to mind much being caught – he is expelled from the commune in 1960.

"I wanted to prove to myself, my parents, my relatives, and my friends in the kibbutz that I was capable of moving forward and succeeding on my own," Naaman later says of this period, "and the proof was money."

So a confused young man, after serving in the air force briefly (he is dismissed for performing aerial acrobatics) decides the best way to salvage his reputation and vindicate his character is to accumulate wealth – but by the very same illicit means that have precipitated his kibbutz exile (he does first work in a bank legitimately – only to be denied a promotion he feels he deserves and then quit in protest, by his own account).

Naaman Diller robs a bank branch in Petach Tikvah of 45,000 Israeli Pounds/Lira (the shekel has not yet been introduced) in 1967. Then the 29-year-old attempts a much more audacious heist in Tel Aviv. He assumes the guise of a utility worker and begins digging up a road – jackhammering the thing – without ever asking permission of anyone or hiding his clamorous burrowing. In fact, he takes breaks from his sweaty labor to sup at Café Vered, whose regulars come to know him as a friendly municipal engineer.

He takes a more extended break from his tunneling when he is called up to army duty at the outset of the Six-Day War – a tour that winds up lasting a full five months. No matter – when his service concludes, he returns to the dug-up street as if he has never paused and presses on with his boring – eventually completing construction of a subterranean tunnel leading to the Tel Aviv Savings and Loan Bank at the corner of Dizengoff and Ben Gurion

Streets. This tunnel he lines with 300 feet – a football field's worth – of pipe.

On the festive first night of the Jewish new year – the holiday of Rosh Hashanah – Naaman Diller approaches his subterranean corridor with a van filled with oxygen tanks. These he hooks up to his pipe. Then he traverses his personal passage, disables the alarm, and cuts through metal bars blocking the windows with a blow-torch fueled in part by the oxygen canisters in the van.

And so Naaman Diller breaks in –by feigning the role of engineer, building out a plumbing system for the transfer of a gas and disabling alarms.

Feeling a certain sense of grime and perhaps hunger pangs, Naaman Diller decides to return home mid-bank heist. He wants a shower, to restore the fastidious appearance in which he takes so much pride (he is said to enjoy wearing suits).

He then returns to rob three vaults and three dozen safe deposit boxes, using further his makeshift torch. And he manages to obtain from these sancta countless currencies and jewels, which he hauls out in bags to his automobile – with no great amount of stealth.

And that should be the end of an extraordinary plot: He has already robbed the vast majority of safes without tripping a single alarm or alerting a single passerby; he has raised no suspicions over the course of months of tunnelling. He has won, essentially. He certainly has no real reason to return to the bank and risk such arousal merely to loot those few remaining boxes whose contents might prove relatively meager. But he has not completed the job in his mind if not every box has been reviewed.

Naaman Diller, oddball and obsessive, returns to the vaults but finds those last few safes particularly difficult to dislodge; again, he does not choose to leave while ahead; instead, he simply yanks and hammers with more force than he earlier applied.

That final bit of boisterous banging is what rouses a neighbor in the dead of night, prompting her to call the police, who catch Naaman Diller relatively easily as the robber walks to his car with yet another bag of loot – they actually shoot him in the knee to arrest his progress. Naaman doesn't mind – and a judge, after psychologists deemed Naaman schizophrenic, hands him only a four-year prison sentence, of which Naaman serves but two years before being set free.

This is a correctional error – for in 1971, an unreformed Naaman Diller breaks into the house of famed painter Reuven Rubin on a Friday night – Shabbos – when the Rubin family is away at its vacation home. Naaman Diller steals 15 canvasses and a great deal of valuables – a traumatic theft from which the painter is said by his son never to recover (his canvasses are like children to him).

An instance of Naaman Diller diligence that again proves his undoing: he saws apart metal bars guarding a window to gain access to the house but so as to throw off the police, he conjoins those same metal rods together with clear tape, making them appear to the naked eye as if they've never been cut. But Naaman similarly restored the metal bars of the Savings Bank when he sallied off for his infamous mid-theft shower. The police recognize the taped-together-grates as the work of Naaman Diller, who is also already known to be quite adept at sliding his slender body between such misshapen rods. Police raid Naaman's apartment – it is rife with property not his own, such that police invite potential victims to pick through the findings. Naaman is sent off to prison again – but not before becoming a folk antihero in the eyes of his ostensible opponents. "He was considered a legend among detectives," says the very cop responsible for this new arrest. "They admired him for his sophistication."[50]

His incarceration this time lasts longer – till 1976 – but also proves more contemplative. No, Naaman Diller does not consider abandoning a life of crime – it's unclear that he even sees his actions as crime, per se. He is obviously impelled by an inner force ordinary citizens can't quite understand. That theft in Tel Aviv of the bank? Naaman went through with it two weeks before his brother was slated to be married – thus dragging the family name through mud in every newspaper before what should have been an unbridled simcha.

No, Naaman begins to consider in this prison stint what valuables he might obtain that he hasn't earlier considered – items less obvious as targets to the ordinary burglar and cop alike. What if Naaman could study antiques – what if he could learn which centuries-old items possess a kind of mechanical magic that render them rare, singular, *precious*?

A fellow inmate whom Naaman queries about such a heist – what do you know about antiques? he literally asks his fellow con – feels the venture too risky and quixotic. Say they did study up on fancy bric-a-brac – how would they ever manage to unload such items to a middleman? You can't bring a Faberge egg – or a Breguet, say – into a pawn shop and expect a) to receive anywhere near its fair value and b) evade imminent arrest.

Naaman, unable to find a willing accomplice, is nevertheless undeterred. But he isn't quite ready either. Upon his release in 1976, he travels to the Netherlands using a fake German passport – his notoriety in Israel having become an impediment to his commission of criminal acts.

But it's not as if his reputation vanishes either – Interpol dispatches a warning to all countries with his photo calling Diller "a dangerous criminal who must be followed."[51]

No legal mind in the world believes this man capable of rehabilitation. For reasons unknown, his recidivist tendency overwhelms all his others – have become his essence. He has been an unrepentant rulebreaker since childhood – and not really out of malice or avarice — just his personal constitution.

"It's something like the vibration of a man next to a beautiful woman when he knows she can be his," he later says, explaining that he never stole for the money. "It's a kind of feeling that you are all alert, focused on the object you want to carry away. You don't even need the object you stole – you just need the excitement."[52]

Predictably, Diller is soon in Holland's custody: he has robbed a jewelry store of goods valued at more than half a million pounds – again, using a parked car, tubes and oxygen-powered thermal burners to get inside.

Once again, he has time in the slammer to ponder exactly which antique pieces are worth having. *Taking.*

***

In 2006, Nicolas Hayek, the man who has pulled myriad Swiss brands from the brink of bankruptcy into a successful conglomerate called The Swatch Group, including Omega and Tissot and Longines, questions in an Associated Press story whether Breguet's Marie Antoinette watch, the piece commissioned to have every watch complication known to horology, even still exists in its original form. It has been 23 years now of radio silence since the theft from Mayer Museum of Islamic Art in Jerusalem.

"I would have put up a $10 million reward," Hayek laments – mind you, his company has just purchased Breguet. "But we had the same issue with the Omega that was the first on the moon. The insurance company says it's their watch. I said it's our watch and the astronaut said it's his watch. If we find the Marie Antoinette, the museum, the insurance company and Breguet will all say, 'It's my watch.'"

But Hayek does not believe he will have to fight anyone for the timepiece because the odds of it coming back whole are slim to none.

"This is a watch that should be treated well. You have to change the oil inside and have the inside checked out to make sure everything inside works."

These quotes will sound ridiculous in just a year's time.

* * *

April 15, 1983. Again Naaman attempts a theft on a Friday night into a Saturday morning – he takes advantage of the Jewish day of rest and makes it his most industrious day of the week. He enters through a back alley – he uses a hydraulic tire jack to pry open the bars of a gate or fence – much the way he has pulled apart bars in previous crimes. He throws a rope ladder with hooks on its end up to a window and climbs 10 feet to its ledge. As he has cased the place, he knows this window will not set off the museum's alarm system, so he merely dislodges it with a screwdriver's end. He slides through an air vent into the museum – he's in.

Naaman spends the night casually taking goods as if programmed mechanically to do so. In the midst of what might be a tense moment even for another accomplished burglar, he eats hummus and leaves the container on the floor among the open cases no longer containing watches. He takes the Breguet 160 but also 105 other watches/clocks and some of Sir David Solomons' oil paintings.

And this time he doesn't make excessive noise – he has ensured quiet by bringing a mattress to land on and clay to muffle sounds. He has brought blankets to wrap up his goods. He doesn't take a break to shower. He does eat plenty, and leaves burger wrappers all over the place. He takes smoke breaks. The Toronto Star calls the unknown party "litterbug thieves." The AP: "The floor of the museum was littered with cigarette butts, soft drink cans and sandwich crumbs."

The strangest thing about the watch mystery involving Naaman Diller is that somehow the perfect heist – one TV station the following morning reporting from the museum on Palmach Street says a quarter of a billion dollars' worth of goods has been snatched – is just another job for him.

Naaman Diller is a professional thief who makes off with watches without a fuss just as he has with lockboxes and paintings. Only this time he has avoided the silliest of mistakes.

This puts the entirety of the blame for the thing on Ohannes Markarian – the museum's curator of watches. He is said to be the only one who understands how they work, how Breguet even constructed them. And the watch thief clearly understood the value of these pieces – which were not widely seen as some tremendous asset to be hidden by triple-thick doors by most criminals.

Poor Ohannes: Now a rotund 50-year-old, glasses perched upon his nose, he is a childhood survivor of the Armenian Genocide in Turkey – after which his family fled when he was 4 to Israel. Ohannes made the most of this escape to the promised land – in Jaffa he learned carpentry and goldsmithing. In the Old City of Jerusalem, he apprenticed himself to watchmakers when he was just 17. And he had a natural talent – a way of absorbing all the intricacies of these mechanisms.[53]

"If Breguet had been reincarnated," the writer John Biggs would later write, "it might have been as Ohannes Markarian."

But now, the authorities are inclined to believe the thief had inside help. So

Ohannes, who has collaborated with the great watchmaker George Daniels in addition to all his other achievements from the time of his youth till now, must undergo a polygraph.

It never occurs to anyone that a man over several stints in prison, a compulsive and recidivist thief, has found himself fixated by Breguet himself. Mind you, earlier in the 1980s, Naaman has gone after an IBM ATM machine having fully studied the actual IBM guide to the machine (Diller spends a whole night trying to crack this money-spitting dispensary only to be caught for having refused to give up his hacking even as morning dawns — for having been compulsive enough to keep trying rather than flee).

He is his own computer. He is programmed. He is a human Breguet – he has innumerable complications and yet he operates in almost a simple, repeated fashion. Again and again. The same MO. Hacking through bars. Oxygen tanks (once he steals them from Ichilov hospital). Eating and smoking and taking leisurely breaks during the most sensitive moments of burgling. A casualness that belies the severity of it all.

And it never occurs to anyone to think about a woman. A woman who meets Naaman at a party when she is just 25 and finds herself utterly smitten with him. She cannot marry him – he is a recidivist thief who will be imprisoned and she will move to Los Angeles and on the eve of her wedding her mother will destroy the mailed correspondence she has been keeping up with him – but for a time they will live together, he will make mango ice cream, they will attend jazz concerts.

Police, watch enthusiasts, Ohannes Markarian in the polygraph get-up being questioned about his associations: none of them understands that time passes very differently depending on your circumstance and like a watch being synchronized with a clock – the Breguet Sympathique mode – if you bring two lives back together, they can regulate each other, become self-referential, align in amplitude and oscillation again.

Yes, Interpol is put on alert and a tip comes into a consul in Zurich that a man named Abu Said may be trying to move the watches. Only a Said never shows up to a meeting he is supposed to attend. England tells Interpol to chase an Austrian named Hubner. In July 1986, a tip comes in from London that in New York, a man from Mozambique has questioned a watchmaker about the repair of three rare pieces. This man is willing to pay $10,000 to get each piece serviced.

Then there are the world-renowned watch collectors whom Interpol investigates – which of them is shady enough to have ordered up such a crime?

And yet there are obvious misses: Naaman Diller is ruled out as a suspect at first because his passport indicates he was out of the country during the heist. Except the passport itself was a forgery, a super-fake, a clone of the real thing. Later, Diller is caught with fraudulent papers in Paris and to pay for his own deportation back to Israel he puts up a rare pocket watch as collateral, engendering not a wee bit of gendarme curiosity as to how the deportee acquired such an item.

When Nili Shamrat's marriage ends in LA, she has the boy the union has yielded and a notion to call Naaman. She does – just to see whether his old number still works. It does. They begin a whirlwind rekindling of their romance from nearly two decades earlier, before any watches were taken, before she left for America. They travel the world and back. Spain, Italy, France, Belgium.

He records their telephone calls when they are apart.

They talk on the telephone every day. And then he tells her he is dying – he has cancer – of the skin, at first, you see, but it has metastasized — to his bones. He indicates it's too late, although he may also decline radiation that can extend his life, however uncomfortably. The year is 2003 – 20 years after the heist.

They marry — knowing full well their sanctified union is perforce a fleeting thing.

As he wastes away, he begins to list all the places around the world (Munich, Basel, Paris) where he has stashed these timepieces he reveals he has had all along. He took them because he was programmed to take – this mechanism does not really extend to resale, to underworld antique-dealing, to making such high-risk larceny profitable.

After his death, Nili will find the watches wrapped up in food wrappers and plastic – maybe wrappers from some of the sandwiches he downed during the theft itself – finds about half of them in total; of those, most are entirely preserved, even if a jewel or two is missing and has likely been pawned (maybe to cover a bail payment or a deportation fee – Naaman never stops robbing).

It's a near-deathbed confession and Naaman Diller passes away as that trope would have it soon afterward. In 2004. Cinematic stuff.

And now Nili is the legal heir to all of Naaman's possessions — the lawful and unlawful of which are perhaps commingled.

Supposedly, Naaman tells her before his death that the statute of limitations has expired. She can honestly and without risk of penalty return the priceless pieces, Antoinette watch included, to the museum from which they were filched.

Except Nili does not quite do this.

Two years after Naaman's death, in 2006, Nili hires a lawyer to sell 40 watches back to the museum for about $40,000. The museum bites – it's happy to have any mechanism returned at all. But international authorities have lots of questions – namely, why wait two years to return the loot Naaman took and never disassembled or discarded (and why charge money, on top of that, and why try to retain your anonymity)?

Interpol, the FBI – they both work the case. And then a break: in 2008, a search of Nili's Tarzana residence uncovers rare clocks and paintings and medieval books — and clues that themselves lead to the lockboxes in which further missing watches are found – bringing the recovered count to 96 of 106 looted pieces. The discovery prompts a classic, horological-existential question: just how in alignment were Nili and Naaman the whole time? Is this the animate, larcenous equivalent of isochronism – of watches placed next to each other adapting one another's rhythms and beats, a phenomenon Breguet studied and much appreciated and incorporated in his Sympathique line of clocks that could regulate the watches stored inside them? Did they marry as star-crossed lovers or so that she could become his legal inheritor? Can you love a watch thief and really not know?

According to the police, she was whisked by the frail, wasted-away Naaman directly after their marriage to the Mayer Museum, where Naaman, a shadow of his former self, nevertheless proceeded to explain precisely how he'd pulled off the greatest theft horology had ever known (or ever will).[54]

Which likely means, in the end, this is a story about a man who was a mechanism — who had to reenact his taking even as his body failed him — and the relationship he had with the one woman who humanized him – who brought his Breguet body to more fleshy life. It's a story with a related, inverse mystery also: Just how much did his kleptomaniacal movement, in turn, impress its patterns upon her?

The Breguet watches are programmed from the start to work via gears and escapements – Naaman is no less robotic in his tunneling into banks and galleries, his slithering inside, his removal of things. The man is compelled, time and again – despite any wishes he might have of turning away from crime. Could that compulsion not but wear off?

The Marie Antoinette was in object form the accumulation of all the watch intellectual property in history to that point: And Naaman took it because he was himself a Breguet mechanism, a rote machine of a man programmed like a module to take.

The great French watch IP was filched by a wind-up clock of a person, whose unwinding had only one release – snatching.

But the police make an even greater find, perhaps, in their 2008 search of Nili's residence: not just the directions to extant absent watches but the blueprints and scrupulous notes Naaman took and then passed onto Nili on his death bed for the upkeep of the clocks and watches – diligent guidance on how to keep each watch running perfectly.

Naaman was a mere thief by inner compulsion – but only at first. In gaining the watches, he obtained also an obsessive desire to preserve the pieces. The watch thief became himself a watchmaker – such was the awesome power of the Breguet apparatuses he stole.

In the Antoinette, Naaman acquired all of the world's horological knowledge. He clearly felt that burden – he understood his role now was to preserve such scholarship-via-object for posterity.

Six years after Naaman Diller's death, a California State Court finds Nili

Shamrat guilty of receiving stolen property – she is sentenced to six years' probation and 300 hours of community service.

But the way this story of the great Antoinette watch full of all the world's horological IP leads to Nili is bizarre and love-suffused and comes with a kicker that so defies a general journalistic objectivity, just basic odds, it leaves one wondering what else do we not understand about the universe. What magical force animates us – this powerful tech Sir David Solomons, electricity wizard, studied? Forget isochronism between two timepieces or the human equivalents – what physical force magically binds us all?

Because that boy Nili Shamrat had with another man while apart from Naaman? Who grew up in California and never dealt with Naaman in any way nor had any relation whatsoever to his transfer of the watches on his deathbed?

By total fluke, owing to awesome, inexplicable fate, that boy grows up into a man who for his own reasons, loves watches – so much so that he launches one of the Internet's first – and longest-running – blogs on the subject. Again, totally independent of Naaman's dealings. aBlogtoWatch.com is almost 20 years old now. Ariel Adams, son of Nili Shamrat, began it based on an intrinsically-felt passion for horology. He is still writing today about the timepieces that daily come out and try to measure up even slightly against the Breguets that passed through his mother's hands on their return to exhibition.

And yet his watch-writing has never been inspired by his mother's tale or that of her thieving lover's – ever.

After I speak to him for more than an hour, he invites me to be a guest on his blog's watch podcast – very kindly. Coincidence – when the timing of two things improbably matches – is a miraculous thing.

We months later catch up over dinner at a fancy seafood place in Lincoln Square. LA-based Ariel is in Gotham for a Citizen event (among others). That night he takes pains to emphasize that his mother may perhaps have been out of her depth when presented with the bequest of stolen watches from Naaman – but if she made mistakes when presented with the stolen property (including the most valuable watch of all time), it was out of shock, surprise, even, say, ignorance of the implications of such receipt. But she never acted out of malice or avarice. Not for a moment.

In the end, Naaman's reflexive theft of the Antoinette watch, Breguet's No. 160, epitomizes the larger business, only very literally:

Time is an enterprise of unbidden taking. And always will be.

Breguet, genius that he was, actually knew this.

Back in his day, he is the first horologist to anticipate truly counterfeiting and theft, though his greatest piece will prove subject to the greatest watch heist ever regardless. Just after the Revolution, in 1795 or 1796, when he could have recoiled more from the terror of the guillotine than from super-forgers, Breguet realizes presciently that clones are both the bane of the watchmaking world and its ever-present parasitical/symbiotic organism.

His solution: take a diamond tipped pantograph – a tracing device with a mechanical arm – and etch, on an almost microscopic scale, into an already-fired-up enamel dial his own name. Literally incise the surface of the dial in a way that would seem desecration were it anyone else's idea or work. Scratch up the topmost layer.

This becomes known as the secret signature of Breguet and it makes him the first to articulate in his own productions the intertwined nature of legitimate and illegitimate enterprise. Now, in 2025, as I write this book, Breguet, celebrating the brand's 250th anniversary, buys a pantograph once owned by George Daniels – the watchmaker who wrote the most comprehensive book (with Ohannes Markarian) on the Mayer collection of watches, including the Antoinette 160, stolen by Naaman Diller. They use the machine to re-insert the secret signature into a commemorative sesquicentennial watch.

The watch world evolves lyrically – or, like the hands its produces, perhaps, merely revolves, ends up where it has already been.

* * *

Oh, and one final note, per horological writer's Carlos Torres' March 2026 archival research findings:

It's eminently possible the legend of Breguet's 160 is itself a fraud — not a watch created at once with every possible complication built in, but a simple gold timekeeper constructed in 1783 that was then embellished by other watchmakers over the next 44 years and, remaining nevertheless obscure, was hyped up only by Sir David Lionel Salomons a century later as being this ingenious creation of a single man for royalty...instead of an accretion of parts, the accumulation of the work of many hands.[55]

The manifest essence of watch IP is and has long been a con.

3

Pre-Westernized Japan, 1860

A story of the Swiss meeting the Japanese.

The journey begins with the death of a watchmaker-merchant in the Neuchâtel mountains named Henri-Francois Perregaux in 1847. This man is about 50, but when he departs, he leaves not only his widow but six children to survive him, three boys and three girls. One is Francois Perregaux of just 13.

The tragedy, as a looming heavy-heeled boot might insects in their carefully constructed conical hill, has the effect of spreading the siblings across the globe, ostensibly in a quest to keep the deceased man's lifework going. Six years later, at just 19, Francois travels to New York, which is by perhaps no coincidence that annum showcasing technologies from around the world, in the Exhibition of the Industry of all Nations (this was the second ever such global showcase and as we've seen, such demonstrations are intrinsic to the development, illicit or otherwise, of horology).

Francois is not only just 19. He's 5'3" (Swiss height will start rising rapidly beginning in those born in the 1870s, in a couple decades) with chestnut hair and chestnut eyebrows and a chestnut beard hanging from his pale, oval face (per his passport).

He would seem unequal to the new world on its face: What business has this gingery mountain boy setting up shop in the globe's ascendant financial locus? What sort of success can he make of an enterprise then learnt at a father's knees when his own had been gone for six years?

And yet the records don't lie: Francois manages to set up a New York City branch for the distribution and retail of his family's watches – H. Perregaux & Co., which he runs for six years, while one of his brothers, Henri, stays back in Le Locle, managing affairs from the home base and keeping tabs on yet one more satellite office, this one even further abroad, in Calcutta.

He sets up office at 78 Nassau Street in what today would be called the Financial District – he is five blocks north of Wall Street, slightly more than two blocks south of Park Row (called Newspaper Row, for its concentration of tabloid and broadsheet offices – by one count, 60 papers are churned out here, soon including the New York Times, once it moves in 1858) and about five blocks south of City Hall.

Young Francois establishes himself in the heart of the Gotham hurly-burly. And while it's not entirely clear how he manages this watch shop amidst the churl – at one point, in 1855, he attempts to send a young watchmaker to Providence, RI, perhaps to work in conjunction with an established New England brand – the evidence suggests Francois was not untouched by the metropolis' darker elements.

Wednesday, June 25, 1856: The start of another sweaty summer in the city. Sometime after Perregaux leaves his facility, burglars emerge from the darkness, like rats scurrying through City Hall Park.

And so, thieves break into H. Perregaux and steal a dark blue leather box – 13" x 11" – containing *59 gold and silver watches from back home in Switzerland and another 16 watch movements.* This is the sort of development that might bankrupt a man without a family behind him, however distant that support might be. Still, the loss must be rectified – so Francois puts an ad in the paper – The Daily Herald, which is printed just steps away. He lists the serial numbers of the missing watches, the escapement type, how many jewels were used as bearings in the movements (to reduce friction between components). He specifies whether the gold used in any given piece is 14 or 16 or 18 karats. He announces a reward of $300 for anyone who can effectuate the timekeepers' return – or about $11,434.45 in the present day, accounting for inflation.

Perhaps the most intriguing horological aspect to the ad is piece #10,871, which seemingly ran two separate movements simultaneously to measure time in multiple places.

LOST AND FOUND.

CAUTION TO WATCH DEALERS, PAWNBROKERS and others.—$300 reward.—Stolen from the office of H. Perregaux & Co., 78 Nassau street, New York, on the 25th inst., a dark blue leather box, thirteen by eleven inches, containing fifty-nine gold and silver watches, and sixteen watch movements, having nearly all of them the name of the above firm engraved on them, the description of which is given below, viz:—

10,810, gold hunting, 18 Rt., white dial, straight line lever.
10,805, " " "
10,421, " " "
10,856, " " "
10,966, " 18 Rt., " lever, 13 jewels.
10,335.36 " " "
10,024, } " 16 R., " lever, 13 jewels, full
10,786, } engraved.
10,025, } " 16 R., " 13 jewels.
10,606, }
11,879, } gold hunting, 14 R., engraved, white dial, lever.
11,348, }
10,518, 10,470, 10,789, hunting, 14 Rt., brass caps, lev., 13 jew.
10,960, 10,791, " 16 Rt., " "
10,944, open face, gold, 18 Rt., straight line lever.
25,061, hunting, 16 Rt., patent lever, full jewelled.
25,823, " 14 " "
24,913, " 14 " "
21,720, " 18 " "
A large hunting, 18 Rt., duplex; winds up by motion.
7,630, hunting, 16 Rt., chronometer; no name on the cap.
11,057, " 14 Rt., lever, brass cap, gold dial.
12,701, open face, gold and cap, 18 Rt., gold dial, lever.
1,664, 1,666, 1,669, hunting, gold, 12 Rt., brass cap, white dial.
16 open faces, gold, 18 Rt., brass cap levers, gilt dials, Nos. 10,108, 10,455, 15,486, 11,275, 11,273, 10,113, 11,333, 11,118, 12,018, 11,342, 10,481, 10,482, 11,130, 12,004, 15,489, 11,315.
10,871, double time, independent second movement.
10,947, 10,811, } straight line lever, comp. bal. movement.
11,129, }
10,820, 10,821, straight lever, ordinary movement.
11,516, 11,518, lever, 13 jewels, red eys. "
11,533, 11,538, } lever, " ordinary movement.
11,067, }
11,878, brass cap, lever movement.
Four lever movements, in tin boxes.
11,989, hunting, silver, brass cap, triple timer.
3,019, " " silver cap, 13 jewels.
11,852, 10,093, 10,094, }
11,303, 11,307, } hunting, silver, brass caps, 13 jewels.
11,166, 11,168, }
11,334, 11,336, hunting, silver, brass caps, 13 jew., (A. Perrot.)
11,018, 11,022, gold hunters, 14 Rt., cassine, full engraved.

The reward of $300 will be paid to any one bringing the above imekeepers to our office.

H. PERREGAUX & CO., 78 Nassau street.

Meanwhile, as Francois Perregaux pleads for the return of gold and silver from the gangs of New York, his relations back in Neuchâtel are deeply engaged in sanctifying an equally dramatic commercial union, if one also ostensibly romantic:

Francois' 22-year-old sister Marie is marrying a star 21-year-old watchmaker-businessman named Constant Girard, who makes timepieces already under his name – the product of which coupling is not only their altered social status as a unit before law and god but the birth of a new watch company to which both sides of the wedding aisle are expected to contribute in equal measure (though it will be a few years before Henri Perregaux asks his mother's permission to liquidate fully the company his dead father and her dead husband bequeathed them).

No more H. Perregaux watch products. No products at all save for those that bear the hyphenated moniker – Girard-Perregaux – a move perhaps most dramatically symbolized by Henri Perregaux trekking to New York in 1858 to sail the seas with his brother.

They depart this loud, industrializing, nouveau riche cosmopolis together, on a steamship called the Asia that pulls out of New York harbor with all the added drama their changed circumstances can lend it. They sold their father's watches as kids and then those with Henri's name on the cases as teenagers. That is done.

Aboard the ship with them – bound for Liverpool on the far side of the Atlantic – are a small who's who of New York society – the young Emily Julia Pierrepont (of Pierrepont Manor)– also in her early 20s (though she will wind up outliving everyone else aboard and die in 1923 at 87) and JL Riker Jr. Also on the vessel – the US consul to Trieste of the Austro-Hungarian Empire (a city only slightly less vital to the Habsburg regime than Vienna, Budapest and Prague) – one Stephen Remak.

Francois and Henri are perhaps oblivious to these dazzling American personalities flittering all about them – or maybe they find a gaiety that requires no translation and enjoy themselves along the voyage.

What's certain, as the ship passes its first night beneath the glinting stars, far away from the dense urbanity that seems to blot out nature itself, is that

Francois, though just 25, has completed fully a chapter of his life.

Less apparent until the boat finally does dock and the brothers make their way south back to the snowy valleys and crags of their canton youth – to Le Locle, Neuchâtel – is that New York was actually a training ground for Francois. For the boy-become-a-man returns to a hometown altered not only by the union of his family with the Girards of La Chaux-de-Fonds but to a Le Locle newly connected to that sister mountainside hamlet by train.

And finally, to greater formal cooperation – bordering on cartel territory – between all the above parties, such that there now exists a *Union Horlogere,* a watch association, whose goal is the exportation and promotion abroad of the products not just from the Girard-Perregaux concern but from all the humble pocket-watch producers of the mountainside.

In fact, by the time Francois returns from New York, it's quite possible the Union already has a very specific goal. In the early 1800s, these Swiss watchmakers doing quite a bit of business in China but far less in the area surrounding it – specifically, in the archipelago of Japan, which was shuttered to all foreign merchants until the US broke that strange, centuries-long blockade in 1853.

Now, the watchmakers saw a nation striving to figure out what it had missed out on culturally during its isolationist years, what elements of exchange it truly desired amidst great upheaval. The Swiss watchmakers had their mark.

There is no record of the watchmaking union – of bearded, calloused elders who toiled on their *far*ms in the summer and assembled watches by the potbelly furnace come blistering-cold winter – formally asking Francois whether, in the spring of 1859, having been home no more than four or five months, he actually wanted to undertake a journey to the Far East (and to be fair, it was about as remote and far a place as their minds could then imagine – a cross-section of Asia to which Western merchants had been denied for hundreds of years – far implies the vacuum of real knowledge they were filling in with their venal merchant dreams).

Francois is still just 25. His mother Rosalie is yet alive here in Le Locle. He has a new brother-in-law in Constant Girard (not to mention Constant's elder brother Numa) – all of whom work for Girard-Perregaux. Why would a man who just came home assent to leave nearly immediately – and this time for a port so much farther abroad – for Yokohama?

But maybe that's just it – everyone he knows and loves is now engaged to labor on behalf of a union whose coming together he missed – what does Francois know of Girard-Perregaux except that, given his surname and his sister's marriage, he's expected to benefit it? Francois was hard at work near Wall Street, near the New York Times, near City Hall – in America. He has never sold watches except those signed by his father or brother – by blood kin. Does he view Constant as an interloper – does he return home only to realize the home he has left is essentially gone?

In any event, the Watchmaking Union decides initially that the 25-year-old should be their second-in-command at this planned Yokohama depot – he shall assist the eminent Rudolf Lindau, himself just 30 but scion of an eminent Prussian Jewish family (that converted to Protestantism). Rudolf has studied in all the best schools, has written for newspapers and magazines and gotten the attention of the wider political world. The Swiss decide this is their way in to Japan. An upper-class polyglot of great erudition and connection and a New York-bred hustler. A street urchin of high horology.

A grand trip begins.

The pair travels first to Marseille in the South of France, then by ship to Alexandria, then by train to Suez (the canal not yet having been finished). From Suez and those narrow waterways below it – the Red Sea, the Bab al Mandeb — the pair sails until they're finally clear of Aden and the Arabian Peninsula entire – until they make it out into the broad Arabian Sea. They stop over in Bombay and then it's on to the Dutch fortress city of Galle on the coast of Ceylon, through the Straits of Malacca alongside of the Indonesian island of Sumatra. A break in Singapore and then finally a ship up to Yokohama, Japan (but first, Saigon, Hong Kong, Shanghai).

Lindau and Perregaux actually split up in Singapore initially – recall that Lindau is a Prussian who speaks many tongues. Lindau has the connections politically to move within the Japanese world that Francois Perregaux presently lacks. So Perregaux stays back while arrangements are made in Japan for his entry; Switzerland, having no treaty at all with Japan, some other country has to sponsor Francois and take responsibility for whatever he plans to do. But this proves an opportunity. While that matter is settled – ultimately, the French delegation pledges their backing to Francois – Francois sets up a Swiss watch office in Singapore.

He finally arrives on the great waves of Kanagawa in December of 1860. Just one month earlier, all those friends and associates he'd made in New York had gone to the polls to vote in Abraham Lincoln – perhaps if Perregaux had stayed on, he would have undergone naturalization and voted the same.

Instead, the Watchmaking Union back in Switzerland has assigned another functionary to Perregaux's crew – Edouard Schnell, a Dutchman. The Dutch were the lone trading partners the Japanese truly trusted, so it's the Dutchman who secures in his own name, but secretly on behalf of the Swiss cartel, land in Yokohama for the Watchmaking Union's offices. Mind you, this neighborhood is basically built on stilts in marshland drained of its natural water by a series of canals (again, a very Dutch way of building but hardly one that inspires great confidence in the structures).

Everything is set up for Francois to succeed – except all those circumstances and market conditions and even systems of measurement the Swiss union back home never considered when dispatching a young man practically to the Antipodes.

For one, the Japanese use a different system to keep time than the Western sexagesimal one – that is, a Base 60 way of counting – unfolded, told and retold, in the endless circles of the Neuchâtel-made pocket watches.

The Japanese divide every day and night into periods of six equal hours each, and those periods wax and wane – vary – as the actual days and nights do. A summer daylight hour obviously exceeds in length a winter daylight hour. Summer night hours run much shorter than their winter counterparts. This system, while highly sensible for an agrarian country whose nascent markets still run dawn-to-dusk, requires clocks both more complicated than those small timekeepers Francois can offer and more simplistic structurally – more imaginative in their conception of time's passage and more demanding of their users, of those keeping actually keeping tabs on this progression.

The imagination is obvious – much like the timekeeping system specified by the Jewish Talmud – they don't bind the hour to numbers that invariably lie by the season – a winter day is to the eye, mind and body an evanescent, depressingly fleeting thing – why should its hours run the same length as summer's expansive, languid periods?

But there the sophistication runs into practical difficulty: These Japanese clocks generally use an escapement – a regulator of the conversion of potential energy into kinetic energy – older even than the pendulum for measuring time – they use two mechanical bars hung with weights – bars called "foliots" (from the French word for "madness"), one for day and one for night. And these are hardly self-correcting regulators, so they must be recalibrated every two weeks as the lengths of hours shift.

Meanwhile, not only do the Japanese have no practical use for Francois' offered timekeepers, they hardly have the funds to buy such a device one-per-person regardless: in a letter back home, Francois and his team write that a single pocket watch costs about 10 to 12.5 times a person's entire wardrobe here. Unlike in China decades earlier, the watches they are peddling aren't luxuries within reach.

Also, the entire foreign enclave in Yokohama is constantly rattled by earthquakes, which usually stir up fires, which can burn down all the wooden housing of the area. This happens repeatedly. And in 1874, Francois is robbed again: this time of a barometer, a clock, books, a gold pocket watch bearing his own name and other items.

Perhaps not unexpectedly, then, the entire Eastern venture is cancelled rather quickly: The Watchmaking Union shuts down its Asian enterprise in July 1863. Two years later, back home, the Union itself ceases to function. In just 11 years' time, another cartel will send spies to America to report back on the best horological advances made in Waltham, Massachusetts.

But Francois Perregaux, essentially stranded in an ever-changing, highly tumultuous Japanese port, does not return home for these malign episodes in the insidious advance of ostensibly Swiss horology. Some part of him has decided that his fate forever after lies in Japan – whether it be fruitful or forlorn.

In Japan, newly open to trade, he will flourish or he will fail, but that setting will have been his choice – after a lifetime of shuttling to various continents on behalf of watch magnates in Neuchâtel.

If he should die lonely and apart here, it will be because he condemned himself to a kind of self-exile (and he is rather alone, at least as compatriots go; history records 150 British and American traders in Yokohama in 1863 but just eight Swiss merchants, Francois included).

Or perhaps he has a family that remains a secret to history even now – perhaps he has married a Japanese woman but omitted his courtship of her from the letters he dispatches back home. Maybe he is staying for a continuation of the great familial-horological line of Perregaux – one lost to history even now.

In 1865, the year the Civil War concludes in his former American homebase, he sets up a company to continue importing and selling Swiss watches in Japan, including those signed by his brother-in-law under the Girard-Perregaux marque. He also retails foreign jewelry and, eventually, carbonated drinks ("Eau de Seltz") and even absinth.

And Francois does not shun the remnants of that union on whose whims he sailed across the globe: He helps collect sociological material, including photographs, for the studies of Aime Humbert-Droz, formerly the president of the Watchmaking Union and now a man keenly interested in documenting Japan on the precipice of modernity, a book about which he publishes in 1870 – "Le Japon Illustré". Three years later, in 1873, slightly too tardy for the Swiss visionaries of earlier decades but momentously just the same, the Japanese adapt Western timekeeping tradition. A GP watch is now more than just a novelty trinket. It is a tool.

We know the Japanese prefer those watches whose cases open up to reveal the movement – they like to see inside the timepiece, into its workings – they hardly want the gears sealed away. They prefer silver cases to gold. They don't like a case that covers up the dial such that one has to release a cover just to gaze at the time. This is an impediment. They considered engravings in the case – dedications or inscriptions of any kind – a defilement of the object, now almost venerated. So the cases are sold clean of the markings so commonly found in the west. The dials are enamel, the hourly markers roman numerals.

The following is an ad Francois places in The Japan Gazette on January 1, 1877.

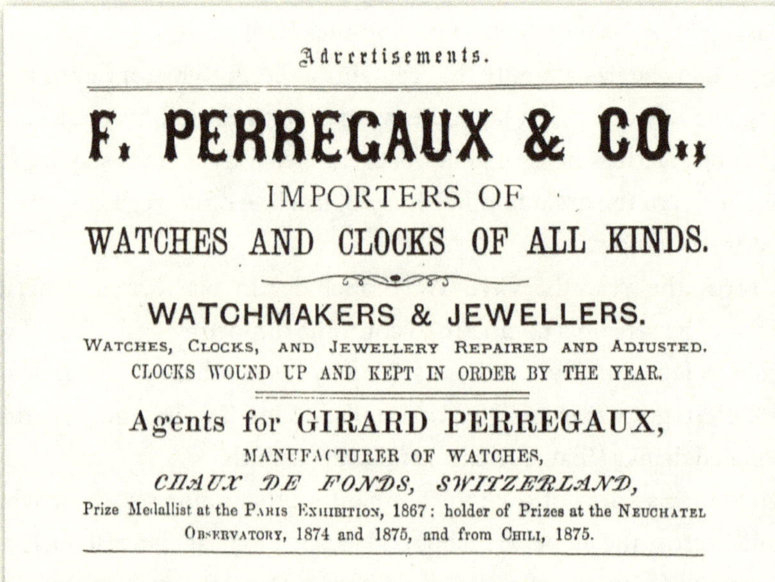

This ad, arranged four years into the new era of timekeeping, marks a moment seemingly bountiful and fortunate for a retailer of watches such as Francois – but in retrospect, it is a monument morbid and somber, for Francois will not live even to see the entire year.

He undergoes on December 10, 1877, what we'd today call a stroke, and just eight days later, he passes. He is buried in the Foreign General Cemetery of Yokohama, beneath a stone still standing today and, enigmatically, next to the grave of a young girl named Eliza who died in 1864, 13 years earlier. A daughter whose premature death drove Francois to stay there forever?

The following March and April bring auctions of Francois' effects, though his Swiss family bans a similar sale of his drinks factory.

And it is perhaps this ban – a rejoinder to the Swiss executors of Francois' Japanese estate published in newspapers – that is the most salient element of his posthumous affairs – for it will be echoed in the coming decades by refusals similarly public and staunch issued by alpine watchmakers in contravention of another Swiss migrants' wishes.

One obituary of Francois published in a French periodical for expats in Yokohama states he was "as loved as he was esteemed." Another adds he enjoyed the "affection of all."

But in the greater history of horological IP, Francois' essential openness to the world, that goodness that wins him such a reputation, is a decidedly secondary feature (though we owe it to the man to record it).

It's the manner with which he is encouraged time and again to sell across the globe watches made in the Swiss mountains by timepiece potentates and then the abrupt way those same *machers*, for reasons unclear, disallow the sale of his seltzer factory that deserves primary notice. That is mirrored in the next push into the Japanese market in the following century.

The Swiss will not give up on Japan altogether – and it's in the great contrast between how the Watchmaking Union treats Francois and how its later incarnations regard his more defiant successor that one sees how horological IP is truly treated by the Swiss.

They have snuck into their mountainous terrain the watchmaking secrets of the Americans; they have displaced in the public mind the Frenchman Breguet as the progenitor of the superlative modern timepiece; they now want to build a fortress and moat around this loot, parapets and fences, too. And the way to do it is for all the companies to conspire together – to collude against outsiders – to do the Watchmaking Union again – but this time with greater devotion and ruthlessness.

Eighty-eight and one-half years after Francois' passing, in June of 1966, the United States Tariff Commission will deliver a report to the President that confirms the Swiss have been working as a kind of racketeering gang for the better part of the prior half-century.[56] The Swiss "had combined and conspired to discourage, restrict, and eliminate the manufacture of watches... in the United States," the report reads.

But we will see first this element of brusque protectionism not in the country Francois first moved to – America – but in the country he ultimately chose as his home – in Japan.

There the cartel will wage its first real war against a watchmaker bold enough to challenge its scheming.

Remember: The Japanese are to be considered as consumers only.

* * *

The armistice is signed, the Great War is over, and across Europe the wristwatch has become a commercial product – born of necessity in the trenches, where a soldier needed both hands free and his time visible at a glance. Swiss watchmakers are growing ever more fearful of a new trend in manufacturing designed to overcome the rampant protectionism gumming up global commerce:

Knockdown production (as it's called in English by historian Pierre Yves-Donzé) – the dispatch of components to factories abroad instead of nearly-finished or even whole items, such that tariffs can be avoided on the finished product for it is not yet existent when the border is crossed. The foreign factory's only real job in this affair is to take those constituent parts and assemble them.

But this comes with tremendous risk, the Swiss believe, even if it allows for tariff evasion – for anyone putting puzzle pieces together might quickly glean the overall picture, enabling him or her to make a product, such as a watch, fully — alone, domestically, with no commitment to the Swiss at all.

The Swiss fear birthing their own future competitors. It's like an industrial Oedipus Rex-type fear – or to be oddly anachronistic about it, it's Anakin in full knowledge that should he bear children, the younger Skywalker will rise up against him (yes, in this analogy, the Swiss are a prescient Darth Vader).

So the watchmakers begin locking the door behind them. In boardrooms across Bern, Basel and the Jura – in those limestone valleys where the River Suze cuts through gorges so narrow the afternoon light barely reaches the factory floor, where spruce forests climb to ridgelines and the snow stays deep well into April – men in dark suits draft the architecture of a monopoly. The mountains have always kept people in as much as they've kept people out; now the industry means to replicate that geography in law.

In 1928, three trade groups signed a pact called the "Watchmaking Conventions." One represented the watchmakers themselves (the Swiss Federation of Watch Manufacturers' Association). Another controlled the movement makers – Ebauches SA, technically a company unto itself. The third united every other parts-maker – cases, bracelets, crowns, you name it (the Union of Watch-Related Branches). Together, they covered every link in the chain from raw metal to ticking wrist.

The terms were blunt. Do business only with one another. Do not sell dismantled watches abroad from 1931 on – except to Germany and France, whom they could ostensibly trust (recall that the two official Swiss languages are German and French). And the cruelest clause of all: after one year, the club was sealed. No new watch fabricator of any kind would be permitted entry.

By 1929, it was done. History will have ended, to paraphrase Fukuyama – mind the year, that of the great Wall Street Crash – and suddenly the group does seem, at least in small degree, quite economically prescient and prudential indeed.

Six thousand miles from those snowbound valleys, in a port city where the air smells of salt and machine oil rather than spruce and limestone, a man from Neuchâtel has been watching the door he came through slowly swing shut.

His name is Rodolphe Schmid. Born in 1871 in the same canton that produced Francois Perregaux – the same narrow valleys along the same River Suze, the same watchmaking families hunched over the same benches – six years before Francois passed away in Japan from "apoplexy," or a stroke, Schmid had entered a world already steeped in horology. He set up a plant in his native canton – the Cassardes Watch Company – and no less spurred on to explore the world, although he seems to have been more independent-minded than Francois, Schmid took the watches he had made in Switzerland around the globe to Japan – to the same Yokohama port community in which Francois was so liked until his untimely death.

Schmid made the journey at the same youthful stage Francois Perregaux did – in 1894, when Schmid was just 23 years old. When he arrived, he found a world that had been dominated by Francois's merchant buddies but one that was clearly about to change forever.

The import of Swiss watches by Francois' Western expat buddies and contemporaries – James Favre-Brandt, the firm Siber & Brennwald and Colomb & Cie – had been the main and only source of fine timepieces on the market. Yet the numbers showed the trend was unsustainable: the rise of the Japanese railway network and the general regulation imposed upon industry by the ascendant, fascistic Empire meant that more folks needed more precise watches. In 1880, only 47,000 watches were imported into Japan. By 1900, the number was 145,000 units.

At some point, clearly, a domestic manufacturer would have to rise to meet the ever-growing demand (indeed, this is partly responsible for the concurrent birth of Seiko by the Hattori family).

On top of that, as Rodolphe Schmid settled into Yokohama, the world's tariff wars were only heating up – such that by 1906, 12 years into his Japanese enterprise, new tariffs made importation of full watches from Switzerland into Japan impossibly expensive – his margins all but disappeared. A decade earlier, in 1895, he was Neuchâtel's third largest watch builder at 20,000 annual units – but that was still a relative pittance compared to the great Japanese demand.

And so two years later, in 1908, Rodolphe began importing not full watches but their constituent parts to a small workshop in Yokohama, where the puzzle was reconstructed, where timekeeping tools were united with their original cases (thus bypassing tariffs). This was a success on a purely economic level – Schmid was able to hire 30 workers and open a new facility for production in Tokyo. In 1910, he made a watch for the Imperial Army – the *Gunjin Tokei*. And in 1913, he was able to begin producing the cases themselves in Tokyo (though based on the same Swiss designs, of course; previously, he sourced cases from his own family venture in Neuchâtel entitled Jobin & Cie.).

On a diplomatic and interpersonal level, however, the venture earned poor Rodolphe nothing but vitriol and antipathy and scorn from his compatriots back in Neuchâtel. They were scared, indignant – and especially intolerant of the new work of case construction. Their great fear? The moment they allow assembly and production to take place outside Switzerland, the Japanese will learn the tricks of the trade – will ascertain how to construct their own watches and assembly lines. And the Japanese are to be customers – and only customers. What if they should learn from the Swiss the way the Swiss have blatantly copied the Americans in decades prior?

So the Neuchâtel Chamber of Commerce wrote to the Swiss federal government of Schmid's venture – of the "selfishness of such an approach, as well as the grave harm that Swiss citizens, who have learned the trade in *our* country, can do to our watchmaking industry."

The cartel went further, officially asking the Swiss government's delegations in Asia "to avoid doing anything to facilitate [Schmid's] business activities."

And it was not just the Neuchâtel watchmakers who tried to bring Schmid down: In 1915, the Federation of Manufacturers of Silver Watch Cases also wrote to the feds, declaring "our national industry stands at a crossroads; watchmaking is threatened by exodus" (of casemaking from Switzerland proper).

By 1928, Schmid had been building his enterprise in Japan for 34 years – more than half his life spent between Yokohama and Tokyo, a man caught between two worlds. And now, back in the Jura, in those same valleys where he'd once learned the trade by the light of a potbelly furnace, the men who stayed behind were signing the Conventions that would formalize everything they'd been doing to him informally for decades.

The door was shut. And Schmid was on the wrong side of it.

Mind you – the Conventions were uniting entities that themselves were already bunched-up holdings. Back in 1927, the American consul to Basel – a man who'd also be posted to New Zealand and whose brother would become the mayor of Savannah, Georgia – reported back to his compatriots on

Ebauches SA with a note of alarm: of 25 movement factories, 90 to 95 percent were joining up. The movement makers were tired of battling on price, tired of producing too many movements for too few buyers – overcapacity that only depressed prices further. So the Swiss Bank Corporation of Basel invested in the enterprise, and its president became the president of Ebauches, too – a banker now dictating the actions of movement-makers.[57]

The message was clear. Three conglomerates and their manifold subsidiaries would be tied formally to each other – to the exclusion of all potential challengers – via the Watchmaking Conventions.

The Swiss government decides this collusive maneuver meaningfully protects a major national industry – and because the groups have taken out loans to promulgate this unified entity, the government also deems the collaborative entity too big to fail – too levered generally and too reliant for those loans on Swiss banks themselves.

So the government of Switzerland actually codified the Watchmaking Conventions in a series of decrees issued from 1934 to 1936. A nascent cartel now had the backing – legally and financially – of a major (*always neutral*) European nation.

The restrictions landed like hammer blows. No timepiece group, however large or small, could start a new watch brand or even expand an existing one without federal permission. It could not hire more employees. It could not relocate plants. It could not consider expansion abroad. Every move required federal approval.

And then the most astonishing decree of all: Swiss watchmaking had been the province of farmers in the mountains who needed a source of income during the colder months – men who assembled escapements by potbelly furnaces while snow piled against their doors in those same Jura hamlets where the craft had taken root two centuries earlier. In October of 1936, the government told those craftsmen they could not tinker inside their own homes. The next year, 1937, all the above laws were consolidated and renewed.

They remained in place, continually renewed every few years, until...*1961*.

In Yokohama, Rodolphe Schmid did not need a gazette to understand what these decrees meant. He had lived their logic for two decades before they were written down.

When, in 1928, he was asked to sign the Watchmaking Conventions – to submit his enterprise to the cartel's authority, to accept its terms and its boundaries – Schmid refused. He was the man who said no. After years of harassment, after letters from chambers of commerce and federations of case-makers, after his own government's diplomats were instructed to shun him, Schmid had an obvious choice: double down on his Japanese bet, on the idea of local construction of timepieces, or bow to the wishes of his paranoid, invidious and hypocritical comrades back home. If there was ever any doubt as to which way Rodolphe might lean, the refusal settled it.

The great historian of horology, Pierre Yves Donzé, has written: "The institutional measures adopted in Switzerland are a rare successful example of state intervention to protect technology transfer."[58]

Successful, that is, if you are on the inside of this exclusive club. But what of those on the outside looking in?

* * *

But those three signatories, all formed in the 1920s, did not wield equivalent power – in fact, none of them was the true powerbroker at all. The real muscle arrived in 1931, when an actual for-profit company swooped in to buy two of the three outright. The historians Donzé and Johann Boillat call what followed the "cartellisation" of the Swiss watch industry. The company took a German name – Allgemeine Schweizerische Uhrenindustrie AG – but everyone called it ASUAG. If that name means nothing to you, its successor will: Swatch – current owner of Mido, Certina, Longines, Tissot, Omega, Breguet and Blancpain.

Now consider the stranglehold. The government threw its financial and legal support behind ASUAG, and suddenly every Swiss watchmaker was required by law to buy movements and parts from ASUAG's subsidiaries. Every single one. The lone exception: companies already using their own in-house movements – and even they were allowed this leniency only with the proviso they buy materials for those movements from ASUAG. A monopoly crafted to print money.

Consider what this meant for Schmid – a man whose entire business in Japan depended on Swiss supply lines. Every movement, every spring, every balance wheel he imported to his Yokohama workshop now flowed through a single entity that viewed him not as a customer but as a traitor. ASUAG could starve him out without lifting a finger – just by enforcing its own rules. The mountains of the Jura had become, in corporate form, a wall around the world.

Two different issues of Swiss publications — L'Impartial of March 4, 1924 and Journal de Geneve of October 3, 1927 – had already featured great Swiss distress at the idea of Japanese watchmaking. And Schmid would hardly bc the last Swiss merchant whose simple method of chablonnage – of assembling Swiss parts locally – would freak out the conservative cartel back home. In the 1930s, responding to another venture called the Tavannes Watch Company, Rolex wrote to the Swiss government: "Once a movement factory is established in Japan by a Swiss company as large as the one in question, we can say goodbye forever to Swiss articles for the East and Far East." Here, it's probably worth noting that during Japan's 1980s economic boom, it was the locus of Swiss watch purchasing, and Rolex sold several million units.

And if all of this seems an obvious violation of anti-trust law, know that such laws don't exist on the Swiss books at this time (in fact, the very notion that competitors should not be allowed to collaborate is rather foreign to the Swiss watchmaking magnate).

Here's how one American newspaper describes the difference in national monopoly statutes during America's highly protectionist, inter-war period – a period when the cooperation prevalent on our shores would be considered in the same polity illegal now:

"In America, of course, the sales associations bring about their chief benefits through gentleman's agreements on price factors, market conditions and such things. In Switzerland, where they have no anti-trust laws, they can go about such matters *more directly* by forming a holding company."[59]

(Some historians assert consolidation is necessary for Swiss horology to survive the Great Depression. And that's partly true: Swiss watch exports fall from 20.8 million pieces in 1929 to 8.2 million in 1932. On the other hand, by 1937, the Swiss produce a full 50 percent of all watches in the world – a 10 percent rise from 1914 and one that comes almost entirely at the expense of American watchmakers, whose global output drops from nearly 33 percent to nearly 25).[60]

ASUAG is a goliath when all is said and done – the world's most efficient maker of watch parts of every kind, bar none. But every Goliath needs a David.

And in Yokohama, the David had already made his move. Two years after refusing the Conventions, in 1930, Rodolphe took his watch company and merged it with a failing enterprise called Shokosha begun by a wannabe local jeweler and watchmaker – Kamekichi Yamasaki.

Their combined new company? Citizen Watch – yes, *that* Citizen – whose naming, per company lore, derives from a suggestion made years earlier by Yamasaki's pal, the mayor of Tokyo, in that the company would provide a watch for every person, low or high (a radically progressive notion at the time; the timeline is blurry here, as Rodolphe started using the brand name Citizen in 1918, years before he partnered with Yamasaki).

All the same, in a very Oedipal way, by attempting to flee a fate, the Swiss watchmakers all but ensured it – they didn't want to see knockdown production morph into a full-on Japanese watchmaker but their stubbornness and lack of grace essentially dared Rodolphe into such defiance.

And the company grew with Rodolphe's insubordinate ambition: it went from employing 30 in 1913 to 110 in 1920 to 200 by 1928. Strangely, Rodolphe only technically owned 1.5 percent of Citizen (Japan still had rather xenophobic laws on the books about foreign direct investment), but managed to control the venture through his proxy loyalists in management, most especially Shinji Nakajima (the longtime marketing director of

Rodolphe's ventures, he personally owned more than 12 percent of Citizen).

The Japanese are only to be consumers: So thought the arrogant Swiss and in the end, perhaps they are proven right in their fear of global mimicry, even if they have it coming, or maybe the following fate could have been avoided if only they'd been slightly less closed off, more welcoming to merchants with differing ideas on commerce:

The first three wristwatches produced by Citizen? All slightly altered takes on Mido (that is, Swiss) models…

But maybe there's a lesson in the value of globalism from that twist alone – for Mido, founded only in 1918, so quite recently, was named for the Spanish phrase "I measure" – "Yo mido."

What the Swiss call their own is never quite just that – never.

Meanwhile, for all its boisterous backlash, the Swiss cartel – febrile and unthinking – focused its energy on the wrong Japanese enterprise – one the company less likely to knock it off its perch. In 1914, Japan produced just 100,000 watches. By 1940, that stat was 1.7 million. A full ninety percent early on (down only to 70 percent by the late 1930s) was the work not of Citizen but of Seiko, of an enterprise begun by the local Hattori clan.

In fact, in 1935, Seiko (aka Seikosha and Hattori) was the largest watch factory in the world.

There is something sweet and retributive – if not ineluctable and obvious – that Seiko, too, depended at first on the expertise the Swiss claimed as their own – Seiko took models from Longines and Ulysses Nardin to study as it designed wristwatches – and it also imported machining tools from the Swiss – without ever having a Swiss merchant in the c-suite. It also studied springs from the Swiss before breaking away and designing their very own (it never occurred to the Swiss that ingenuity existed in Japan such that their own technologies, once evaluated, could likely be topped instead of merely replicated).

The Swiss took for study models from Longines – as in the company helmed earlier by Ernest Francillon, the most prominent advocate of espionage of the American Watch Company in Waltham.

Karma.

But what we see from the divergent reactions to Francois Perregaux and Rodolphe Schmid is the Swiss mindset of the era in action: anyone can buy their watches for a price – but heaven forfend they should take an interest in watchmaking itself. The latter is verboten – for everyone but them.

And they have a boardroom of bankers and a federal government to back up that protectionist (and slightly xenophobic) stance.

* * *

And yet, it is a position they will soon alter – but only in America and for no stated reason. The Swiss will ostensibly now work with the watchmakers of another country without fear of the incidental transfer, let alone theft, of IP.

Why? What is the reason the Swiss suddenly feel comfortable revealing their methods? What do they know about the US market that even those who regulate that industry in that market don't?

The Swiss return to the US 60 years after spying there – with what schemes, notions, assurances?

4

The Swiss Mob and America, 1936

The year is 1934, and Franklin Delano Roosevelt is trying to coax the world back into commerce after years of Depression, tariffs, and clenched national fists. Washington is opening drawers, unfolding treaties, talking again in the language of reciprocity. On paper, at least, the idea is one of thaw.

This nixing of ruinous tariffs is accomplished via bilateral treaty – with 19 different countries, including Canada, France, the United Kingdom and, of course, Switzerland – whose value of trade with the US has dropped from $94 million in 1929 to less than $35 million under the Smoot-Hawley duties.

The Swiss treaty, the tenth of the Roosevelt era, arrives with details so strange they border on comic. Signed in January 1936, with provisions taking effect in February and May, it requires Switzerland to buy 90 percent of its lard from the United States. Ninety percent. Of its lard. In return, America cuts tariffs on Gruyere and Emmenthaler by 29 percent. The agreement reads less like geopolitics than like two bickering lovers haggling in the pantry so as to avoid having a far more difficult discussion in the bedroom.

The key component to the whole deal is the watch.

Of 59 categories of product in which the treaty drops rates, a third of them involve watch parts. For Switzerland has decided it cannot push around the US as it has Japan, and perhaps it's to its own economic benefit not to try

– Switzerland finally *wants* to ship watch movements for use in US cases and timepieces (what about the US figuring out how to make Swiss watches and stealing the IP? What about the risks of knockdown production? Just wait.).[61]

And the US thinks it has itself a fantastic deal, even though a cursory overview of the watch issue should make negotiators think twice.

The American negotiators know the surface facts. They know that many watches sold as American now depend on Swiss guts, especially Swiss movements. They know the domestic brands that once boasted of self-sufficiency are increasingly casing someone else's machinery. What they do not do is sit with that humiliation long enough for the obvious question to darken the room: how did the country that led the world in 1876 wind up begging the Jura for its innards?

Why should the US need watch mechanisms and parts when it was the world leader in 1876, on the country's 100th birthday?

The US trade reps never consider this.

And so a treaty is signed – and not just that, but a section is further appended that conscripts the US to halt the importation of nearly a million counterfeit Swiss movements annually (again, does anyone wonder how the Swiss have suddenly become so desirable that there even exists a black market for its movements?).

What Washington learns first is not the full mechanism of the cheat but the style of it. The Swiss do not honor the treaty with plain dealing. They work it with charm, with anxious phone calls, with flattery and procedural fog — the manners of a man smiling in your doorway while sliding his shoe over the threshold.

In October 1936, as the Swiss franc is devalued against the dollar, the Swiss minister in charge of trade frantically calls his counterparts in the US at least twice in one week, begging the US not to exit their pact via its escape clause (which allows for the raising of duties if one country or the other is found to be dumping products in the other's market against the spirit and letter of the deal or otherwise manipulating currencies/trade).

A telegram fires from Washington to Switzerland. Wilbur J. Carr, irritated by the Swiss barrage of pleading calls and in no mood to soothe a government that keeps testing the edges of the deal, decides to put the message in writing and send it straight across the Atlantic.

Wilbur J. Carr is bothered indeed by these nonstop, panicky Swiss calls asking the US for guarantees it won't exit the treaty, when it's the Swiss who keep pushing the boundaries of the permissible.

The cable is addressed to Hugh Robert Wilson in Bern, a diplomat with the sort of résumé that already feels faintly cinematic in retrospect — Yale, Wolf's Head, the polished discretion of the foreign service, and ahead of him, though neither man can yet know its full shape, a posting as America's last ambassador to Hitler's Germany before the continent goes to war. Perhaps he reads Carr's message at a broad desk in a room gone dim in the Alpine afternoon, perhaps in a properly tailored suit, the sort of man accustomed to receiving history in typed bursts and expected not to flinch.

Cables Carr to Wilson: I've told the Swiss that the US "Government could not bind itself as regards future action" though it has "no present intention of invoking the escape clause and that it probably would not do so." All that said, *you* should lean on them, insinuate they need to fall in line:

> *You may also intimate that pressure may be brought to bear upon this Government (the US) to abrogate the Trade Agreement in view of the impetus to Swiss exports which may result from devaluation of the Swiss franc and that* this Government will be in a much better position to withstand such pressure *if it can be shown that the Swiss Government has taken measures to facilitate the entry of American products into Switzerland...*
>
> *The Department desires that you follow the situation very closely and insist that in accordance with our trade agreement, American trade receive a fair share of all advantages resulting from any relaxation of Swiss import restrictions.*[62]

That the Swiss take advantage of this treaty in the following decade and half in ways the US will only come understand later is the reason the next agreement between the nations doesn't occur until 1951 – and deals only with avoiding double taxation on estates and inheritances, with ensuring no citizen of either country faces levies by both.

The old American houses can feel the theft before they can diagram it. Hamilton, Elgin, Waltham — companies that still want to make watches with their own movements — watch the market grow and feel almost none of the effects. The numbers do not make sense. Demand is there; reward is not. Somewhere inside the machinery of trade, somebody is manipulating commerce, and they begin pleading with Harry Truman to look harder.

On the other side of the table sit the companies that have already capitulated and now rely on Swiss parts – Bulova, Benrus, Gruen – firms that firmly reject the idea of additional duties. They enlist a former senator named Millard Tydings to advocate on their behalf (Tydings, rather honorably and uniquely, attempted to pass legislation condemning Hitler's antisemitism from the Senate as early as 1934).

Now, in one sense, Harry Truman is a watch man. When he takes the oath in the war's closing stretch, a Gallet Flying Officer is on his wrist — a Swiss chronograph crowded with information, its rotating 12-hour bezel and inner city ring built for a world suddenly organized via flight. It is the sort of watch that makes its wearer feel both worldly and grounded at once.

Of course, Gallet is a Swiss brand (it was just recently acquired by Breitling, incidentally, as I write this), founded in 1826. The watch, a gift from two of Truman's former Senate staffers, is the product of a nation he is supposed to be scrutinizing.

Truman is a watch man but seemingly one who knows but half the story.

In 1952, the US Tariff Commission endorses a 50 percent rise in duties on Swiss timepiece imports. Truman rebuffs that advisory group, rebuffs Waltham and its fellow US watchmakers seeking to remain fully American.

He waves the argument away. To raise the tariff, he says, would be to indulge a fantasy, to pretend that 'serious injury exists when the domestic industry fails to gain something it never had.' It is an elegant sentence, brisk and managerial. It is also the sort of sentence a man can utter only if the deeper history has been successfully hidden from him.

What Truman fails to understand or cite is that the American companies indeed once possessed such a dominance of the market – though it may predate his governing days, it seems almost shameful that even the American president cannot recall the time, not long ago, when his compatriots were leaders of this industry. When Waltham ruled the world.

No matter – Truman is celebrated by a certain elite pundit for his position – and then by the very Swiss companies alleged to be manipulative. The Times says his decision-making is "in the best tradition of American trade diplomacy."[63] The Times further reports that "many a glass of cool white wine was drunk today" in Geneva.

The Journal de Geneve writes: "It is not an exaggeration to say that the decision taken yester-day by President Truman has a tremendous importance in reaffirming the confidence between his country and the free nations of Europe."[64]

The party does not last. Two years later, President Eisenhower goes along with the Tariff Commission and raises the watch duty by 50 percent over the levels codified in 1936 (this is perhaps an unexpected turn; only weeks earlier Eisenhower has ignored his tariff commission, refusing to raise duties on a new cod and flounder product called "fish sticks").[65]

And yet the truth is still hidden – why exactly aren't American companies competitive in a booming post-war economy – why does their share of the pie seem rather meager?

The factory floors tell the story in bodies. In 1951, 12,000 American workers produce 3.2 million watches. Three years later, half of them are gone – 8,000 workers grinding out 1.6 million pieces, the production lines thinning like hair on a 40-year-old. Meanwhile, the companies that could be considered subsidiary to Swiss production employ at least 15,000 American workers – nearly double the domestic count. And while American watch production doubles in volume during the 1950s and 1960s, its share of the global market ends up halved.

Why?

And how is it that the purely American firms can produce and retail a 17-jewel movement at cheapest for $10.50 while the Swiss are able to create the same thing for just $4?[66]

The answer comes in October of 1954, or three months after Eisenhower imposes the tariff. And it's an answer anyone familiar with the Watchmaking Conventions of 1928 in Switzerland could have guessed at, only no one did. But before we reach the belated US discovery, a word on how the Swiss managed to insinuate themselves into American industry such that they were able easily to eviscerate it from the inside.

What was really going on behind the scenes?

* * *

From the time the Swiss timekeepers form a cartel in the late 1920s, those watchmakers realize they need enforcers to keep their constituent companies and their rivals in line (the Swiss – never quite so neutral as they claim). They need the Swiss equivalent of heavies, in other words – which is to say, wily, clever, and uncompromising business minds. A very specific kind of business-world tough.

There are two such goons, if we can use that term without being totally cruel.

The first is Sydney de Coulon, born in 1889, a man who will live many decades more – not dying until 1976, by which point he'll have already eviscerated American manufacturing in his chosen horological sector, and with no apparent compunction. He wears fine, three-piece suits, dark-rimmed glasses and slicked-back (though heavily receded) hair. His eyes lock in on those he meets in an unnerving way, unless you are used to his very Swiss, matter-of-fact demeanor. His forehead seems to stretch out into the distance like a gradual Alpine slope. Or a bunny hill at a ski resort.

He is cold, strict, a stickler for rules (though the rules are his own so he fears not running afoul of them). And for all these traits, he is appointed director of *Ebauches* SA (one of three entities that sign the Watchmaking Conventions to inaugurate the cartel and a firm that builds not-fully-assembled watch movements – this is the definition of an *ebauche* – and provides them to brands).

De Coulon is also a figure in ASUAG, the super-holding company that owns Ebauches SA (and the precursor, decades earlier, to the Swatch Group). He is the first secretary of the ASUAG board and subsequently its vice president. Historian Johann Boillat will later call De Coulon "the personification of the Cartel," such is his fierce devotion to the cause.[67] The great historian of the watch industry Professor Pierre Yves Donzé, a business history professor at the University of Osaka, in Japan, writes of Sydney De Coulon:

"He is the man who personifies the cartel and the Statut horloger. He embodies the Swiss watch industry which successfully took on the American challenge of industrialization and then organized itself within rigid structures."[68]

Sydney de Coulon is said to have a keen sense of the watch business and also a sense of humor about his role, yet the record impresses upon one the picture of an ambitious man who pursues what he wants without smile or any other outward sign of good spirit. He is an enforcer of the cartel, after all – and one who truly considers the Swiss watch industry a national cash cow that must be safeguarded from all those who'd try overtaking it. Hence his effort, as Donzé notes, to box America into rigid confines – but more on that in a moment.

The other heavy is Maurice Vaucher – a former Church pastor with a degree in theology, of all things – and the president of the Fédération Horlogère, the watchmaking industry trade group that itself is a signatory to the Conventions establishing the cartel. Vaucher takes the position in 1933 and will hold it for 24 years, until 1957.

He is the face of this operation – the lobbyist gallivanting about – he does not toil in the shadows. He attends trade fairs and expositions – always nattily attired – and when the cartel needs to deputize a fresh face to argue on its behalf in Washington, DC – when the cartel must beg, borrow and plead to avert a US tariff increase on Swiss watches, Maurice Vaucher is the man sent to do the job.

Vaucher is also something of a wordsmith or trickster – when the US administration levies a charge of market manipulation at his Swiss constituents, Vaucher turns it right back on 'em: whatever moves Swiss watchmakers are undertaking are all in the name of quality control. Nothing nefarious to see here. We Swiss are focused only on maintaining the very best production lines.

In fact, if you feed all of Vaucher's letters (and subsequent trial documents) into a proper AI, as I have, that program will tell you this excuse is but one thing: "regulatory capture as a brand defense."

Or BS, in the name of taking over the entire watch industry.

But language is one thing – how do Coulon and Vaucher, the timekeepers' thugs and sweet-talkers, actually seize the watch world from its diverse participants – its manufacturers across the world – but in America, most especially?

It is a classic bait-and-switch – or perhaps better put: An unmerciful rug-pull.

First, in the 1920s and 1930s, the Swiss make an offer the American brands of great repute cannot refuse: Wristwatches and their mechanical technologies are developing quickly by leaps and bounds; the Swiss companies offer to provide replacement parts for a new generation of American watches at a very low cost. The American watchmakers cannot resist such good deals on parts that allow them to revolutionize their wares and their servicing operations (building parts such as these on one's own is expensive, by comparison). The Swiss – de Coulon and Vaucher – are offering a tremendous deal.

And then in 1931 the Swiss begin to make demands. The signing of the Collective Convention in Switzerland means watchmakers there cannot export any parts without adhering to new cartel rules. In 1934, the Watchmaking Statute, a Swiss law, forbids horological enterprises from increasing production or installing new factory-floor apparatuses without governmental approval.

And here comes the Rug Pull: The Swiss companies repped by de Coulon and Vaucher tell the American watchmakers that have become reliant, like drug addicts, on Swiss parts (such that they've stripped away part of their own manufacturing capacities in America) that the Swiss laws now apply to all American recipients of Swiss parts.

Suddenly, the Watchmaking Federation is setting the prices on American watches and their production numbers, too. And the Swiss have set up the "Watchmakers of Switzerland Information Center" in the US in order to observe which American brands are compliant and which shall be banned from ever receiving Swiss parts again – even if those companies no longer can create the parts in their own facilities.

Meanwhile, de Coulon and Vaucher both go on to become politicians – de Coulon serves on the Swiss Council of States and Vaucher on the Swiss National Council.

Their very obviously antitrust tactics are rewarded – by an electorate heavily reliant on the money made from watch production (and the safeguarding of gold and the confection-concocting of chocolates).

Which leaves American companies in a tight, if not impossible, spot: They have for decades procured parts by giving up to the Swiss cartels all decision-making power – what and how much they can make, where they can make it (so the size, scope and location of their assemblages), where and to whom they can sell it and for how much.

How they may or may not acquire tools of repair for these watches they are assembling even.

Dangling on the edge of a cliff, the US watchmakers, while not saved, are, at the very least, given a legitimate hearing in Washington – the Federal government suddenly interested indeed in this Swiss rug-pull that has left US manufacturers adrift and bamboozled.

October 1954. A federal grand jury in lower Manhattan hands down an indictment charging 18 American companies and six Swiss firms with violating the Sherman Anti-Trust Act and the Wilson Tariff Act. Time magazine breaks down the charges – and the specifics are devastating:

> *Bulova and Gruen "in exchange for Swiss movements and parts, agreed to limit their US manufacturing, import from no country except Switzerland, permit a cartel-hired accountant to audit their books."*[69]
>
> *Benrus Watch "agreed in 1945 to give up manufacturing watches and*

> *parts in the US and to liquidate its plant so nobody else could make watches there."*
>
> *"Three Swiss makers agreed to set up no plant in the US, while the subsidiaries of seven others allocated markets, restricted re-exports from the US, agreed to buy only from Switzerland."*[70]

Agreed to give up manufacturing. Agreed to liquidate its plant. Agreed to buy only from Switzerland. The word "agreed" doing extraordinary labor in those sentences – as if a man handing over his wallet at knifepoint has "agreed" to part with his cash.

And then there is the advertising agency of Foote, Cone & Belding, which is employed by two of three signatories to the Watchmaking Conventions of 1928 (which formalized the cartel operations of Swiss watchmakers). The *ad agency* is accused in the indictment of actually policing American watchmakers – keeping them in line, ensuring the companies aren't producing any watches in excess of what the Swiss have deemed their upper limit – on behalf of those two signatories (Ebauches and the Fédération Horlogère).

And though this suit is filed at the end of 1954, it covers actions going back to 1931 – that is, since even before the striking of the 1936 Trade Deal the Swiss minister was so desperate to keep intact via his phone calls to the State Department.

Bulova, Benrus, Gruen, Longines-Wittnauer – all technically American firms using Swiss components – are charged. As are the American subsidiaries of the big boys – Rolex, Patek – and even the aforementioned components conglomerate and Conventions signatory Ebauches SA (which itself is owned by an even bigger monopolist – the super-holding company ASUAG).

It thusly begins to make sense: The Swiss have not feared IP theft in America as they have in Japan because they've so constrained the nature of the work US companies do with Swiss parts, they've so circumscribed the US companies' decision-making, nothing has been left to chance, let alone the free hand of the market.

The agreements the Swiss struck forbade the American companies from creating anything of their own – even from exporting finished goods to third countries. The Swiss have extracted every usual US corporate prerogative in exchange for its tech (which they themselves stole decades earlier).

The breathtaking scope of what the Swiss cartels have been demanding of these affiliate companies, these corporations turned vassals, is astonishing. As the Times sums it up:[71]

> *The complaint said the defendants had agreed...to fix the prices and terms and conditions of sale for Swiss watches imported into the United States; to blacklist, boycott or fine American watch companies that did not adhere to these agreements.*[72]

To make matters worse, all this time when American tariff-dons were debating, for decades, the extent to which Swiss movements were being fairly shipped into the US, they were overlooking the most obvious and salient point:

The Swiss offices of these American companies – that of Bulova, that of Gruen – "were signatories to the 1928 Conventions and members of the Swiss watchmaking cartel."[73] Unbelievably, Bulova's Swiss office was invested in by the Cantonal Bank of Berne, while the Swiss office of Gruen struck a deal in 1943 to export 300,000 Swiss movements out of Switzerland in return for Gruen limiting the type of movements the company made in America to just *two*.[74]

That is, put more simply: To import Swiss parts, the US companies have killed off their own engineering capacity and trading know-how. The Swiss even jointly arrived at a price for repair parts from Switzerland – so even the watchmaking necessary to keep a mechanical timepiece ticking is tied to collusive pricing (at the expense of watchmakers and customers).

Which means that even if a court case now suddenly finds the Swiss liable for such a scheme, it is perhaps already too late for American watchmaking on an industrial level to be salvaged – the knowledge and facilities are lost – the US companies have already been coerced into letting go that basic acumen such industrial production requires, the atrophy is impossible to overcome, the learnings reabsorbed.

Many of the US companies have been hollowed out such that they'd struggle figuring out how to be fully independent once again, anyhow.

The antitrust case is a classic example of the call com*ing* from inside the house – of a truth so staggeringly obvious it is bound to be overlooked. The Department of Justice states in its case that the value of yearly watch sales in the US is $225 million – and yet 75 percent of the fabrication of those watches takes place in Switzerland. It's a staggering fraction that should have raised alarm bells earlier.

The Swiss companies are seemingly lenient with their American partners because they aren't partners in any real sense of the term – the American officers are puppets.

A look at some of the renowned Swiss defendants from the federal case makes that clear:

Henri Stern Watch Agency, for anyone in the know, is the US distributor of Patek Philippe (just how much rope was Patek ever going to extend its American arm?; Henri Stern, for the uninitiated, is the son of Charles Stern and the nephew of Jean – the latter two of whom supplied dials to Patek before taking over the entire operation).

Jean R. Graef, another defendant, is the official US retailer of Zenith now but studied once upon a time at the storied watchmaking school in the Vallée de Joux, where a watch he constructed in 1933 with two balance wheels (it measured time by averaging out the two balances) was so accomplished it was put on display to inspire future students (to which American entity would he ever concede horological power?).

Then there's the American Rolex Watch Corporation – yet another defendant – whose name speaks for itself. Would the central office of Rolex, the most buttoned-up, conservative of companies, the most centralized, "Severance"-style-Lumen-like of the Swiss timekeepers, the company most known for C-suite meddling in even piddling affairs, ever grant its US arm even the least bit of autonomy?

Perhaps, unsurprisingly the above defendants are among 11 (of a final count of 23, after a 24th defendant is dropped) who choose not to bother with a trial (one sure to illustrate the clear limits of their autonomy and the full extent of the top-down dictates) and to enter into a consent decree, instead, in March 1960 (two non-watch importers also agree to the terms, raising the number of defendants who submit to the new rules to 13).

And if you want to know how controlling and anal and underhanded Rolex can be (and I am asking you to make that judgment, as I would never do so myself; I just present facts), please consider that in 2006, nearly 50 years after Rolex enters into the consent decree, the US Department of Justice alleges the watchmaker is in violation of that agreement — and has been, by nature of its own policies, since 1996 — *for a full decade*.

Specifically, according to the DOJ, Rolex has been banning the insertion of Rolex parts into watches containing non-Rolex accessories (so sorry if you have previously altered your own product or customized it in any way — your watch can't be serviced now, even though you own it); Rolex has been prohibiting the resale of spare parts under any circumstance (so sorry if you wanted to serve, as a trader, the independent market for repairs and maintenance — which are, of course, measures necessary to the long-term survival of any mechanical product); and Rolex has been capping itemized parts markups at 50% (in other words, if your independent watchmaker around the corner buys from Rolex a spare part to service your timepiece, and he invoices you such that he lists separately his cost of labor and his retail price to you of the new part, so sorry but he cannot independently choose a price for the spare part he puts into your watch — Rolex has capped what he can charge for it, thereby interfering in his own profit and loss ledger, precisely the sort of hierarchical tampering the entire consent decree was designed to eliminate).

But this being Rolex, the company manages to get off without admitting wrongdoing by paying a paltry $750,000 to the Department of Justice. But it doesn't just pay the sum. It begs for revocation of the 1960 decree it has allegedly been violating for a decade altogether — implying it should be allowed now to restrict access to parts, to use its vertical integration to deny the right to repair and to force indie watchmakers out of business (technically, the corporate entity Rolex Watch USA files a motion to terminate the consent decree).

Heinously, the DOJ complies. And so Rolex, having ostensibly breached an order it had quite publicly committed to follow, gets off after handing over a paltry sum and somehow having the rule voided itself.

The US government's reasoning to accede to the request? The old judgment "no longer serves the procompetitive purpose it was intended to serve, and its continued existence does not provide a public benefit."[75]

Ask an independent watchmaker who just wants the right to repair a Rolex in his or her shop lawfully and without losing money whether the decree, compelling the fair sale of parts, is superfluous and silly.

The remaining 10 defendants who attempt to beat the rap fail to. The trial plays out before Judge Cashin without a jury over five months in 1961 – Ebauches SA, Bulova, Gruen, Wittnauer all found guilty when the verdict finally lands in January 1964, three years after testimony concludes and two months after JFK is assassinated. The Federation of Swiss Watch Manufacturers' (FH), along with Ebauches another of the three major signatories to the 1928 Conventions, is convicted alongside them.

The strangest thing is the gap – five months of trial in 1961, and then silence, almost as if the verdict is the subject of negotiation and wrangling between national governments, until the ruling surfaces in January 1964.

Judge Cashin's ruling is rather basic – the convicted may no longer restrict the commerce of watch companies with which they do business. And the convicted can't find some third party to restrict the watch companies' work either. Those to whom you sell parts are your clients and not your serfs.

This will be one of Judge Cashin's final monopoly cases – born in 1892, he will retire in a year, after a long legal career that included serving in the prohibition department of the United States Attorney's office in the 1920s, chasing bootleggers. He will pass away in a half-decade, in 1970.

But really, it's not his ruling that sums up best the Swiss plot to insinuate itself into the American horological system and take it over by hollowing it out – like larvae who, left to nest inside a living creature, grow strong by eating their way out.

No, the matter is framed best in a June 1966 Tariff Commission Report on watches presented to President Lyndon Baines Johnson (he's considering the watch tariff issue a final time). This document winds up being the last word on the Swiss monopoly. And it cuts to the core of the Swiss business tactics that have underpinned decades of international commerce (remember: the Japanese are to be treated as customers only – and so are the Americans, as it turns out).

First, the extraordinary thing: the Swiss respondents repeatedly refuse to furnish witnesses. That is, when the commission – the body that can once-and-for-all impose punishing new tariffs – asks the Swiss to come testify, the Swiss simply decline.

And yet, the commission declines to endorse those new duties – because the Swiss brands at issue have already moderated their behavior somewhat since they were brought to court by the Feds.[76]

"Had the order of the US District Court not intervened," the report writes, "the issues before the Commission might have been different."[77]

But all the same it underlines an underhandedness the judges, for the most part, seem to abhor. On the issue of a cartel being legal in Switzerland: "This 'cartel' did not confine itself to regulating trade within the Swiss industry, but extended its restraints to trade and commerce elsewhere, including the United States."[78]

The dagger – the trenchant line that calls the Swiss cheaters who only ceased because they got caught and who seem always to bounce back, resiliently, with some new scheme:

> *Several of the respondents have demonstrated in the past both the ability and inclination to impose unreasonable restraints on trade...Moreover, these same respondents have demonstrated in the intricacy and constant propagation of their commercial* arrangements a remarkable degree of adaptability and inventiveness, and a propensity for using their joint power not only to secure a competitive advantage but also unreasonably to inhibit competition.[79]

Again, the only reason the Tariff Commission decides not to recommend action against the Swiss at this juncture is because as far as it can tell, the companies have reorganized since they were first brought to court a decade earlier in 1954 and because right now, they seem not to be in contravention of section 337 of the 1930 Tariff Act. On that narrow question, the commission must excuse the Swiss watchmakers. But of their historical sliminess – that unethical elasticity when it comes to obeisance to the law – the commissioners are quite certain.

(All of which recalls a non-denial-denial from an attorney for the Swiss watchmakers in 1965 after the president of Hamilton said the Swiss cartel is in the habit of issuing threats to destroy US industry – the lawyer does not deny the Swiss have made such remarks – only that they have not done so recently – that their interference is "ancient history.")[80]

The question for the Swiss, then – having attempted to suppress a Japanese industry (to modest effect) and having been caught doing the same in America but even more insidiously – becomes:[81][82]

What if it encounters an opponent in the watchmaking world – a national foe – wilier and even more amenable to breaking the law than itself? What if its flexibility is made to seem utterly rigid – its operations ponderous – in comparison to some new speedy adversary, one more responsive to the market than it has ever tried to be?

Enter first Japan and Hong Kong with revolutionary timekeeping technology – and then China, with an economic ruthlessness and underworld savvy overwhelming even for the Swiss.

5

Asia Goes its Own Way: Japan, 1930; Hong Kong, 1973

Japan in the late 1960s – it is a rocket-ship on its way up, an Astro Boy polity, a nation-state with anime oomph and manga power – a neon, cybernetic, ghost-in-the-machine skyway Hollywood hasn't even yet imagined.

It's not quite "Blade Runner"– but it's on its way.

The yakuza films of the Nikkatsu studio are all narrow and wide anamorphic shots and bursts of color (their finest protagonist, Jo Shishido, has cheek implants, such that his face puffs out like *fugu*).

Since it hosted the Olympics in 1964, Japan has also been on a building spree – what first began as a project to show outsiders this losing World War II side was resilient has become a self-sustaining exercise in modernist redefinition. In unburdened self-expression.

And the financial center of its capital – the Ginza District of Tokyo – bears all the hallmarks of this Japanese rise – it's both Western and Eastern architecturally, clearly influenced by the forces that once firebombed and occupied the place but also devoted to crafts native and coveted – marked by venerable outlets that long predate the failed Imperial Project and are now propagating a very 20th-century hybrid Japanese vision.

A branch of the department store Mitsukoshi, vendor of high-end kimonos, ceramics and lacquerware, in business since 1673, is nevertheless comprehensible to the foreigner traipsing through its displays; its rival, Matsuya — whose company history reaches back to 1869, though its landmark Ginza store opened in 1925 — spends the mid-1950s and 1960s hosting design exhibitions , shows such as "From Space to Environment" (1966).

And then there is the Ginza headquarters of *Hattori Tokeiten,* or the K. Hattori Company - the watch firm that changed Japanese timekeeping forever.

Kintaro Hattori was born in 1860, the same year Francois Perregaux washed ashore in Yokohama to sell Swiss pocket watches to a country that didn't yet want them. Where Perregaux was the Swiss cartel's emissary, dispatched and then abandoned, Hattori was a local boy sent into his own community for work - the only son of an antiques shop owner, Hattori apprenticed himself (or was loaned out by his father) to a wholesaler of Western goods when he was just 12. But in that job, he had the prophetic epiphany about the watch industry that would inform all his future decisions:

The shop at which he worked had days when nothing sold — rainy days. But, Hattori noticed, a nearby watchmaker, in addition to new sales, had a recurring revenue stream from the upkeep of timepieces already vended — and that repair money served as a rainy-day fund for moments when sales, for whatever reason, flagged.[83]

And so the young Hattori switched jobs. He left the wholesaler Tsujiya for the Kameda Clock Shop in Nihonbashi, and two years later, moved to the Sakata Watch Shop (when the last of these went bankrupt, Hattori gifted his accumulated, saved salary back to the downtrodden proprietor — amazing as that sounds).[84]

Finally, Hattori opened a watch repair shop in the Ginza in 1881, refurbishing and reselling junk pieces. A little more than decade later, in May 1892, he opened a clock factory in a derelict former glass manufacturer that he wishfully branded"Seikosha," which literally means "House of Exquisite Craftsmanship" or "House of Precision Workmanship." He employed a dozen workers, including an engineer he considered his partner in the venture,

Tsuruhiko Yoshikawa. Oddly, the Japanese wouldn't buy pocket watches that were not engraved with various patterns.[85] Apparently Yoshikawa had impressed Hattori with a machine Yoshikawa himself invented by which he could easily engrave pocket watch cases with "nanako," or fish scale patterns. By July 1892, they'd produced 12 pendulum clocks — using primitive cutting machinery powered by a wooden flywheel called a "buriwa" that had to be be rotated manually.

You can still see it — the dozen taking turns putting all their weight against the groaning wooden logs, each one heaving him or herself into the primitive, croaking apparatus, wiping sweat off their brows with towels during breaks. Their panting as they try to catch their breaths. Sipping water, beads of perspiration perched on their upper lips.

Hattori had no great affection for this outmoded way of doing things. A year into his Seikosha venture, he moved his residential office to a bigger space and acquired a five-horsepower steam engine. But he knew that was but a minor upgrade — and so was the 25-horsepower engine that replaced it. The real advances he coveted could be found only abroad. So Hattori took two tours of the world's great watchmakers — at age 39 in 1899 and again in 1906. He visited small workshops in France, Germany and Switzerland, yes. But he took serious note of the industrial operations at the two US giants to which he was granted access (into which there is no record of him sneaking): the watch factories at Waltham and Elgin. "In his visits to the Waltham and Elgin factories in America," reads the current Seiko Museum text, "Kintaro learned rational layouts for standardized mass production lines."[86]

He brought back from the first trip a 60-horsepower engine for his enterprise. On the second, he acquired a 140-horsepower engine.[87] Yoshikawa, the inventive engineer, was so inspired by the machinery that upon his return from one trip he devised and built a new automatic pinion lathe — it raised pinion output (the creation of smaller gears) from 100 pieces per 12-hour day via the operation of 25 workers to 200 pieces in the same amount of time — via the operation of a single worker. The old lathes had about a 30 percent reject rate for their output parts. Yoshikawa's new wonder produced virtually no defective components.[88]

And Seikosha started pumping out its creations: the first pocket watch — the humbly named "Time-Keeper" came out in 1895 and the first wrist watch, The Laurel, in 1913. Its big hit in the 1890s was an alarm clock that didn't rust. By 1901, the plant employed more than 300 people. Big breaks came in 1910, when Seikosha began creating its own hairsprings and in 1913 when it managed its first enamel dials. In 1924, it started using the name Seiko on certain timepieces.

Lore has it that always Hattori was motivated by one particular mantra: *"Isoguna, yasumuna"* — neither rush nor rest — or, Rush not, rest not. An argument for a kind of controlled and constant commitment to the task. For a life lived in accordance with the very motions perpetual-yet-well-pulsed to be found in the mechanical timepieces to which Hattori had devoted his life.

This philosophy was put to the test. The Great Kanto Earthquake of 1923 leveled all 23 Seikosha factory buildings plus the grand commercial space Hattori had built in a former newspaper office in 1895 and famously topped with a clock tower. All was ravaged by fire except a single water tower at the plant and, in some accounts, its perimeter walls.

Hattori's response? He took out a full-page newspaper ad promising to replace every single watch that had been in Seikosha's possession for repairs, all 1,500 that had burned to a crisp, with brand-new pieces, free of charge. It was his duty to the bereft Japanese people.

In 1932, two years before Hattori's death, the company moved to a curved building that bent ever so gently around a Ginza street corner and was topped by an even grander, 50-foot-high clock tower. The building's exterior stones were hauled from Okayama Province for the lower floors and from South Korea for those higher.

The history of Seiko, more or less.

Now, it's the end of the year 1969, and inside this upscale business district astride past and present – thirty-five years after the founder's death but governed still by his creed of rhythmic, perpetual movement– a very different object shimmies with such regularity. A rock.

Seiko engineers and technicians have produced the first quartz watch for commercial consumption – the Astron – which they will put up for sale on Christmas Day.

As an invention, this device does not quite stand alone – Bulova has already introduced an electronic watch, the Accutron, with a vibrating tuning fork whose oscillations replace old-fashioned escapement components and whose battery obviates the need for a wound-up mainspring (whose unwinding releases power in traditional mechanisms). Hamilton has also already introduced a battery-powered watch whose electrical impulses cajole the timekeeping mechanism along (these have batteries instead of springs, too, but they maintain the old-fashioned balance wheel of the escapement).

And this particular quartz watch is more of a demonstration of Japanese ingenuity (or a halo product) than an item for the masses – the Astron, which harnesses a battery to spark quartz to vibrate at the mineral's resonant frequency and then takes those oscillations and, dividing them up, feeds them into a stepper motor, quite literally costs the same as a compact car – 450,000 yen, or $1,250 dollars (more than $11k today with inflation).

More important than the Astron in particular is the technology it conveys to the world – the possibilities it opens up to rivals beyond Switzerland without the IP that country's heavies have for so long rather ingloriously guarded.

Which isn't even to mention that quartz oscillations, outnumbering those of mechanical watch escapements exponentially, are proportionately that much more accurate – they lose or gain seconds by the month instead of day.

(A true story about Japanese dedication to precision: The US had the patent to the atomic clock, which elevates atoms in an electromagnetic field to higher energy states and then measures the frequency of the microwaves emitted as they transition back down to a lower state; the US got the clock working plus-or-minus one second every 650 years; then Seiko acquired the license and reduced that error rate further – to plus-or-minus a second every 1,000 years – as if that infinitesimal difference mattered – though as a matter of pride to the Japanese, it did; the insanely regulated clock actually became part of the 1970 world exposition held in Osaka.)[89]

By 1976, the price has cratered. Seiko releases the Type II and SQ series – quartz timekeepers that cost between 18,000 and 30,000 yen – from about $60 to $101, depending on fluctuations in the exchange rate. The Astron's technology, no longer a curiosity, is now a technology anyone can afford.

Citizen – the Japanese watch power born of Swiss antipathy toward production abroad – a company that with its owner was blacklisted years earlier – is also involved in this new age timekeeping:

Its Citizen Crystron is actually a greater market presence than any single offering from Seiko – Crystrons can be bought for cheaper or for higher rates – depending on whether you want an analog display or a liquid crystal one. As for accuracy? The Swiss don't have a shot with their hoary mechanisms at matching the Crystron Mega – which loses 3 seconds per year (granted, it retails for $15,000 and comes in gold).

On the low end, the United States suddenly can offer watches whose prices, if matched by the Alpine bean counters, will wipe out all Swiss margins. In 1976, an LED digital quartz watch manufactured by Texas Instruments retails first for $19.95 and then $9.95 (back in Japan, the Casio F-Series is similarly cheap). No decent Swiss mechanical watch can be offered at that price – not even close.

Then Hong Kong enters the fray. Quartz technology lowers the barrier to entry, and entrepreneurs in the Special Administrative District set up production lines of cheap little watches no one will ever feel guilty buying – plastic doo-dads that, ironically, keep almost perfect time. By 1978, Hong Kong is the single largest manufacturer of watches in the world by volume.

If the Swiss have indeed finagled away mechanical timekeeping from some rightful pioneer of that tech; if they will soon co-opt the legacy of the ingenious French horologist Breguet; if they have spurned and blacklisted compatriots who simply wanted to build products abroad, such as in Japan; and, finally, if they have collusively rigged the entire US watch market, hollowing it out in the process, such that it can never recover, do they view this quartz moment as their deserved, if delayed, comeuppance?

Likely not. But they know they're in trouble. Seiko has been nipping at their heals since before the introduction of quartz, such that the big push in the late 1960s by Omega and other Swiss brands represented by Gerard Bauer and his cartel — the Fédération Horlogère — is not to get their watches on the wrists of astronauts as those brands might now have you think: It is to get them on the wrists of earthbound Soviets — as many as possible — and soon — before the Japanese conquer the USSR market first.

The secret Russian connection is first formed in 1966, when the Russians visit a Swiss fair for watch machinery and place orders on equipment whose export is no longer restricted by the time the meeting is disclosed in a letter two years later by trade delegate Raymond Probst.[90] The Russians are somewhat aggressive — they even contact the cartel oversight men in Biel — the Fédération Horlogère, which hosts the Russians at a luncheon — not likely out of common courtesy but perhaps. Or the Swiss already see an opportunity to seize a market. The following summer, that of love, 1967, the Soviets invite an FH delegation led by Gerard Bauer to Moscow so that the Swiss can tour the best Russian watch factories.

Their conclusion?

> *"The Russians are currently capable of producing a good, medium-quality watch, but that they are likely still about 10 to 20 years behind the top Swiss standards technically and industrially. Nevertheless, the watch industry in the Soviet Union has developed so rapidly that today, with an annual output of approximately 18 million wrist and pocket watches, it has become the second-largest producer of small watches, trailing Switzerland but ahead of Japan and the USA. In addition, it produces roughly 20 million large clocks."*

There is a market here, in other words, and it's one the Swiss could easily dominate with its superior mechanisms. By the following summer, in July of 1968, the Swiss, led by Bauer — but now under his aegis at the head of another industry body, this one called the Swiss Watch Chamber — have struck a basic framework agreement with the USSR — officially titled "Protocole de collaboration scientifico-technique et économique."[91]

It's a rather loose document resting on three pillars: the Swiss and Soviet horologists are to cooperate on matters scientific and technical — by visiting each other's labs, sharing research and academic articles; they are to assist each other mutually in the development of the opposite party's foreign trade; and they are to respect each side's intellectual property.

While the original impetus for this loose almost-deal is the Russian desire for Swiss technology, the Swiss see the potential accord as a way of forcing the Russians out of the watch counterfeiting game they've played so coyly (and in the process, suppressing the movement of counterfeit watches through Hong Kong) and as perhaps a legitimate economic opportunity to open up the USSR to Swiss watch brands.

In any event, it's hardly a binding document on either party. And the Soviets get plenty of time to mull it over — they aren't slated to sign it till the Swiss make a return visit in September. But that September 1968 visit is postponed when in August — just after the Swiss have departed Russia — the Soviets invade Czechoslovakia.

Writes trade delegate Raymond Probst of Swiss calculations regarding the signing thereafter: "President Bauer is well aware that in the current political-psychological situation such an act cannot be considered."

But it certainly can by the following March of 1969, when a Bulgaria-born Swiss technocrat stationed to the Moscow embassy named Jacques Fardel writes up a report on potential collaborations between the USSR and the Swiss in various industries — chemical, mechanical, horological. The watch leaders want the thing signed — before Japan takes away more of their potential business:

> *The watch industry is very interested in the as-yet unsigned watch agreement of July 1968 and hopes that this framework agreement will soon be signed and come into force. Both the Watch Chamber (President Mr. B. Clerc) and the F.H. (Minister Mr. G. Bauer and Mr. Retornaz), as well as most of the industrialists whom the reporter was able to visit, see in it a possibility to initiate new business in the USSR. In any case, no disadvantage can be seen in this watch agreement for the Swiss watch industry, since the Russians are scientifically equally advanced, but lag*

> *behind in manufacturing technology. This lag cannot simply be caught up through ordinary visits and exchanges of opinion. The danger to the watch industry is Japan!*

The figure spit out is that 40 percent of Russian watch production is defective anyway — they are so far behind in quality control that their scientific expertise is for naught. Amounts to zilch Besides, the Swiss can control quite what they reveal: "the exchange of technical goods can be steered."

There *is* the problem of counterfeiting — the Swiss leaders believe the Soviets export yearly 1.5 million watches with supposedly "Swiss-made" cases (they're made using Swiss machinery) beyond the Communist bloc. But the rush to beat Japan is of greater importance. So Bauer declares — this can't wait any longer — the agreement must be signed this spring (even though it doesn't actually require anyone to do anything). So reports Fardel, the Swiss diplomat — who's also an engineer — writing the status report. Fardel interviews individual brand proprietors on any potential deal, fascinatingly. Perrenoud, the general director of the small brand Doxa in Le Locle "is very positively disposed" and Fardel notes the plant is quite modern, if small.

Portescap and Reno SA, of La Chaux-de-Fonds, is a watch supplier that should very much be opposed. In 1934, Portescap invented and patented the Incabloc shock protection system for balance staffs. The Soviets nevertheless reproduced this component without change — leading Portescap to confiscate the Soviet watches whenever they popped up at exhibitions in Western Europe. But strangely, that IP theft bred a funny familiarity between the crooks and the inventors they copied, and Portescap seemingly has taken that entanglement as an opportunity to sell the Soviets its Vibrograf assessment machine — which can measure a watch's beat and rate — its accuracy. So they're in favor of further deal-making.

Omega, in Biel, tells Fardel, the attache to Moscow, that it's way ahead of the rest of the Swiss, but they've slowed their Eastern efforts, halted them in fact, so they can join the wider team. As early as 1967, Omega, which produces 1.5 million watches per year, visited on its own Moscow, Minsk, and Leningrad — and ever since it has exchanged technical papers with the

Soviets. "OMEGA had the intention of achieving a proper agreement with the USSR for the exchange of know-how as well as a commercial contract," Fardel writes. "This was stopped in view of the watch agreement with Swiss industry. It goes without saying that OMEGA now expects a *swift* signing of this agreement."

Omega tells Fardel they can only sell 100 pieces a year in Russia right now. They expect an agreement would change all that. It must. Soon.

And then there's FAR: Les Fabriques d'Assortiments Réunies in Le Locle. FAR is an essential components supplier that employs 2,000 people, produces 200,000 parts in a year and whose general director says, "Visits by Russians and Japanese to his factory are out of the question, as is any exchange of know-how."[92] FAR in the 1980s will be forced to merge with spring-maker Nivarox due to dire financial circumstances – with Nivarox-FAR eventually being sucked up into the same Swatch Group conglomerate as Omega itself. Meaning all these brands being solicited for varying opinions on an opening with Russia will all wind up a single company.

But for now, the tally makes clear the almost unanimous clamoring for a deal with the USSR.

The deal is never signed. And so the Swiss shift: Maybe Latin America is their next big market. In Argentina, the Swiss attempt in 1969 to get the tariffs on their watches lowered. The Argentines link the issue with Swiss quotas on Argentine wine — accept 10,000-12,000 hectoliters of our *vino,* the Argentines suggest, and we'll take your watches. The Swiss will only go as high as 2,000 hectoliters — and would prefer the wine issue not even be mentioned in the same sentence.[93] They add that the importation of gold watches legally will help Argentina curb precious metal smuggling. Then they raise the global wine acceptance to an international 6,000 hectoliter rate.

By August 1970, it's clear that Latin America is no less complicated than the USSR — maybe more so. In Brazil, the Swiss watch magnate Edgar Kocher has since 1949 run Eska Relogios Ltd. in São Paulo. Now, Citizen, the Japanese watchmaker presently led by Eiichi Yamada, is planning on making a big push into Brazil. Secretly, Bauer and Probst and the other horological diplomats of Switzerland conspire with Kocher to ruin Yamada's impending visit. Kocher plans to launch a press campaign via a group called the Brazilian Industrial Syndicate — whose name and other identifiers betray no connection to watchmakers — to smear Citizen and Yamada or Japanese watchmaking more generally — using material provided by the Swiss. The nature of the campaign or the smear is not made clear even in the formerly confidential Swiss documents I obtained about the plan — the accusations to be leveled maybe relatively unspecific or benign.[94] Maybe it's more fear than smear campaign. But the bottom line is that it's designed to unsettle Brazilians about the presence of the Japanese watchmaker in their country. The Swiss realize there might be a terrible rebound effect, if the result of the anti-Yamada campaign is greater tariffs for all foreign watches or a simple fear of the other — an unleashed wave of nationalism. It's unclear to me the degree to which they follow up and cooperate with Kocher, as a result (I can't find their paper trail, if they do).

In the end, the Swiss search for global markets that can somehow compensate for falling sales in traditional Western ones. The so-called Quartz Crisis proceeds — although it isn't really a technological phenomenon, this near-bankrupting of the Swiss industry. And the feared Japanese are only a small part of the economic story.

As René Retornaz, the deputy to Gerard Bauer at the Fédération Horlogère, will later recall, the calamity is really a series of macroeconomic setbacks and headwinds — compounded and exacerbated by the absence of a necessary supply chain.[95]

The 1970s features an oil shock in the US due to the OPEC embargo — US watch sales take a hit from rising gas prices with no corresponding watch purchasing increase in the Gulf states raking in their petro-dollars; the '70s also feature a collapse in the dollar value against the Swiss franc, so watches become 2.8 times more expensive in the world's most important market simply based on forex; and a sixfold rise in the price of gold from 1972-1975 and finally a price of $800/oz in 1980 render the main input cost for Swiss precious metal watches far too high for the pieces to remain profitable at established prices; finally, Switzerland does indeed lack a supply chain for quartz watch components — which network ultimately will takes the Swiss six years to build out (whereas the Americans already have an inbuilt font of the tech due to its longtime development at both NASA and the Pentagon).

Companies either go dark forever or combine into conglomerates of failed brand names. This is the lowest point for Blancpain and Omega. A nadir of commerce for Universal Geneve, Lemania, Enicar and Cortebert (all of which cease to exist as they once did– with the latter two disappearing from the world altogether and the first, Universal Geneve, just now being revived by Breitling in the year 2025, after becoming a zombie brand owned by a Hong Kong investment concern for years).

The Swiss industry employs 90,000 workers in 1970 and only about 28,000 by the mid-'80s.

Of course, Switzerland survives it – visionary consultant Nicolas Hayek, empowered by creditor local banks to extract value from two groupings of Swiss brands facing insolvency, instead fuses the two together and manages to turn them into a single profitable watchmaking enterprise; Hayek assumes the CEO position at this retrenched and reinforced conglomerate; in 1985 he manages with local investors to buy 51 percent of the company he is running and therefore gain almost total control of it; Hayek finally brings to market a Swiss product that can compete with the cheaper quartz-driven watches that have eaten the Swiss share of the market's low end – the plastic, translucent Swatch.

Books have been penned already about the marketing attendant to all of this – how Hayek and his team – not to mention maverick Jean-Claude Biver – manage to elevate those remaining, solvent Swiss brands into paragons of luxury in the public mind – into synecdochical ideas of the good life they had never quite symbolized before. Mechanical watches were once utilitarian items — no more.

What does get overlooked (because this global push was kept confidential for a period) is that the Quartz Crisis is actually a quest to remake the world — a race against time to push timepieces into protectionist countries run by autocrats that never before seemed likely buyers of Swiss watches.

But when when you export 50 percent fewer watches to the US than you did just years earlier, you don't have the privilege of discounting any areas of the map as economically forbidding backwaters.

Also overlooked — the Swiss invented the electric watch — they were just too otherwise absorbed to make a fuss about it or commercialize it.

Le Locle, Switzerland, February 1932. A retired watchmaker named Georges Emile Pellaton-Steudler, 67 years old, announces the creation of a watch powered by a battery (instead of a spring) he has developed over 12 years – its power regulated for timekeeping by a flywheel, a disc able to conserve energy and therefore help dispense it smoothly.[96]

While electric clocks already exist – and in fact, Bell Laboratories has already begun using quartz crystal in timekeeping in 1927 – Pellaton deserves credit for the anachronistic futurism of his device (the battery can last a year on a full charge) and also for its size – this is a battery-rechargeable *pocket* watch.

Strangely, Pellaton does not seem to care much for the device's commercialization, but then, he is an inventor's inventor, a kind of head-in-the-clouds thinker who enjoys solving problems for the sake of it and will live to the ripe old age of 95 doing just that. (He also may fear the mechanical watch cartel in Geneva that will have to approve his electric watch being placed on the market.)

Anyway, Pellaton follows up this invention with a new, cheaper electric typewriter, just a month later, in March 1932[97] and finally, in October of 1946, with a smaller, more compact electric wristwatch – this last device being perhaps the most useful to humanity and therefore the most egregiously overlooked and un-commercialized of all his work.[98]

Imagine if a great electric wristwatch had emanated from Switzerland in the first post-World War II year. Would the Swiss have developed the quartz tech behind the 1969-released Seiko Astron long before the Japanese company?

6

Was It All About Marx from the Start? Or, a Deeper Look at Longines

What if everything recounted, while being utterly true, is also, if only by omission, thoroughly false?

That spying effort set off by a Jura mountains council of watchmakers, led by the head of Longines, that preys upon a great New England factory in Waltham — is it really entirely explained by industrial worry and transnational competition?

What if there are more profound forces at play here — existential reasons for the surreptitious surveillance?

What if I told you the Longines-led spying effort is in part a reaction to *Karl Marx?*

Not to Marxism, mind you – Ernest Francillon of Longines is reacting to Karl Marx the living irritant, Karl Marx the catalyst, Karl Marx the man whose decisions turn the watchmakers in Francillon's own town into the founders of a movement meant to render owners such as Francillon obsolete.

The movement is anarchism.

* * *

Longines is never an entirely independent enterprise. The company's own

180th-anniversary research concedes that its first two supposedly self-made calibers — the 20A in 1867 and the 18A in 1868 — begin as unfinished 'blanks' supplied by Japy Freres and Robert. So before Francillon ever sends anyone abroad to pry into a rival's methods, his firm already knows how to incorporate the work of others.[99] Of course, at this moment in time, all watchmaking enterprises borrow parts from others and finish them — this horizontal orientation is in part the reason the Swiss are falling so far behind the Americans in the manufacture of timepieces.

The ledgers tell the story. Longines is desperately reliant on the American market, and when American firms like the one in Waltham - the American Watch Company - and the one in Elgin, Illinois - the National Watch Company - begin increasing their quality and stealing market share (via vertical integration), the Longines sales collapse: Revenue in America plunges from 18 million Swiss Francs in 1872 to just 4 million in 1877. Longines retails 15,732 watches in the US in 1874 - and then fewer than 3,000 from 1876-1877. Francillon takes out a 60,000 Franc loan from a bank in Lausanne in April of 1874 and another of 50,000 in November. Longines' US authorized retailer also takes on massive debt.

So Ernest Francillon does has every economic reason to begin a spying campaign of great shadiness, one that can undermine the Waltham factory and perhaps restore Americans' desire for his exports.

But there is a political factor at play in the town Longines occupies — and it's every bit as potent and explosive an issue as Longines' declining export tallies: The watch factory owners of the Jura Valley are at war with their agitating laborers— and vice-versa.

December 1868: Ernest Francillon bans members of Karl Marx's International Workingmen's Association from his workshops — he must prevent workers from organizing, demanding more rights, attempting to seize in any way the means of production. He puts it in writing plainly: "I believe I must announce that henceforth, anyone belonging to the International Workingmen's Association will no longer be admitted to my workshops."[100]

But this ban hardly calms a factory town on edge: St. Imier, home of Longines, is a pressure cooker. The town is full of skilled men being told that men like Francillon are exploiting them for their labor.

In the early 1870s, the watchmakers still number in the hundreds, if not thousands. They are accustomed to a certain historical way of working – to artisanal laboring separate from one another in small mountainside shacks in the winter.

But increasingly, the Swiss watchmakers are being told they will inevitably have to slave away on rationalized factory lines with none of the warmth or tradition to which they are habituated. The Marxists have seen this happen in other fields and professions. The watchmakers must refuse this revolution at all costs. It is a war all but inevitable between worker and owner.

And the advent of the telegraph and news photography mean the workers in Saint-Imier have access to a wider world – they read newspapers dense with foreign dispatches, market intelligence, arguments about wages, alternative payment models that they can import into their little valley. If only they take action.

The remarkable thing is not merely that the workers devour these papers. Their managers read them too. The same sheets that denounce overseers and praise labor are also among the fastest, sharpest guides to the economic world beyond St. Imier — a nineteenth-century Bloomberg terminal printed by men who want to overthrow capitalists.

Meanwhile, the St. Imier workers of Longines feel oppressed even by the village's most basic organizing principles. This tiny city asks of them to keep track of *four different time zones* within it – those factories that do already exist keep time slightly differently than do the official municipal office and the train station and the provincial church.

(Switzerland is not alone in having its times not synced even by the middle of the 19th century; in England, for instance, before the advent of the railroad, in 1830, Oxford was five minutes and two seconds behind London *always*; Exeter was 14 minutes behind; Leeds was six minutes and 10 seconds.[101])

And then Marx makes a move — in late summer, 1872: At congress at the Hague, he argues for a centralized system of socialism to benefit workers. He believes a General Council should be at the center of the seizure of political power from the propertied classes – he wants a singular apparatus overseeing the revolution he has been slowly explicating all these years in the papers.

In other words, he essentially betrays his own teachings, "Animal Farm"-style: All workers are equally deserving of economic control. Only some workers are more equal than others.

The St. Imier watchmakers are furious – they will have none of that – they see centralized power-structures as being all the same, so far as what workers ultimately earn, how they benefit from the scheme – or don't.

Marx has crossed a line.

But they do not repudiate what the revolutionary earlier preached – that their extreme craftsmanship should be recognized and remunerated and not discarded in the coming age of mechanized industry. The St. Imier horologists still abhor the notion of leaving their precious ateliers and working on degrading factory production lines.

Eight days later, on September 15 and 16, 1872, watchmakers and their allies gather in St. Imier and answer Marx: They split away from him. In the very hometown of Longines, among men and women whose fingers are trained to crate hairsprings and escapements, the international anarchist movement is born — born of a refusal to trade one master for another.

The new Anarchism movement is initiated with the passage of four key resolutions by attendees, including such global luminaries as Errico Malatesta, Carlo Cafiero, and Mikhail Bakunin.

That's right – just four years before the head of Longines sends his right-hand-man to spy on American watchmakers, he bears witness to the birth of anarchism in his backyard – to the rise of a movement that, left unchecked, will displace him and every other timekeeping magnate in Switzerland.

The nightmare, from Francillon's side of the desk, is double. The Swiss might lose to the Americans abroad and to their own workers at home. Waltham will take the market, while the literate, radical craftsmen of St. Imier turn the workshop itself into a political experiment in which factory owners are supplanted. In that vision Swiss watchmaking does not merely decline; it loses command of its own story.

This, too, is why Francillon sends his spies.

Consider again 1876. Francillon and his fellow Longines bigwigs (plus the other timekeeping chiefs on the Jura Council) have witnessed revolutionary ferment for four years since the declaration of anarchism as a movement. Of course, they want a leg up on their American competition, however unfairly it's gained. But the juxtaposition of these two events, the fright and panic anarchism has given the most important watch don of St. Imier (birthplace of that movement) and his spying thereafter on the Waltham factory floor in America, home to precisely the protocols anarchists loathe and resist and fight the most, suggests a truth deeper and more expansive than what this book has been peddling until now (by intention).

The surprise kicker:

The head of Longines does not send his technical director David merely to observe US industrial advances – but to gather knowledge so thorough and irrefutably productive that it will help him and his contemporaries maintain an entire social hierarchy – and their position atop it.

This is not just a battle for tech supremacy – it is an existential gambit to retain power.

If the American system of factory watch production can be shown unbeatable and unyielding, how many Saint-Imier or Swiss workers will protest its adoption – how many will stand up and argue for the methods that guarantee their coveted trade's local demise and their collective loss of income?

We cannot know fully what is in the heart of spies and their dispatchers. But we can say this for sure:

What the spies bring back from America is not just technology but a system of corporate power enabled by it. We know that, 150 years later, "Rolex" is a term more commonly associated with the Swiss than "anarchy." Orson Welles infamously calls Switzerland a dull place free of social turbulence in his depraved monologue at the end of "The Third Man."

No, Francillon succeeds — at least in the narrower, harsher sense that matters to men trying to remain on top. His spy brings back not just technique but discipline: a factory logic he can impose on St. Imier, and its labor agitators. A system for taming the restless and the unwilling, for turning grievances into offenses.

It is a city of merely 8 square-miles, after all.

Saint Imier cannot magically yield both the anarchist movement and the great heist of American factory methods by the anarchists' boss *simply by chance* within a few years' time. The revolutionary declarations and shift to American production methods have to be related – the latter a response to the former.

There's also proof of this in Longines' very payrolls from the 1870s. Remember when Francillon declared all members of Marx's International Workingmen's Association persona non grata? He didn't foresee that, in splintering off from Marx to create anarchism, the local watchmakers would slip back into his employ. The historian Florian Eitel has painstakingly spent years cross-referencing the Longines Archives and the Jura Federation (the local anarchist umbrella group) historical collection. He is able to identify that of the 40-50 founding anarchists in St. Imier, a full 14 are also present in the Longines employee register between 1867 and 1880.

"Nearly one-third of the local anarchist leadership was economically tied to the very industrial machine they criticized," Eitel adds.[102]

And the American spying effort – the mechanics of the factory Francillon cribbed from Waltham – has a material effect on these anarchists who have infiltrated his rolls. For instance, before he renders Longines more fully mechanized and vertically integrated, Ernest Francillon very much needs the particular assistance of Ali Eberhardt, a *remonteur*—that is, a final assembler of all the watch's components. And also an avowed anarchist who has snuck in the Longines gates. Ernest Francillon requires the services, too, of the assembler Samuel Schneider – another anarchist. But once the heisted American methodologies take hold of Longines, Francillon has a factory in place of manual assemblers such as Ali Eberhardt and Samuel Schneider. Their services are worth far less. Eberhardt's daily pay slips from 5 to 7 francs in 1872 to 4 to 5 francs in the machine era after the spying. Schneider suffers even crueler reduction: whereas he had once been classified at Longines as a "qualified artist," he is rebranded in the machine age a "semi-skilled machine tender."

Intriguingly, the assemblers at Longines of the balance wheel – what the Swiss who speak German tab the "unrueh" or "unrest" – are usually women in this era – many of whom are married to anarchist leaders. "Through familial networks," Eitel writes, "the anarchist movement effectively 'infiltrated' the most delicate assembly departments."

Obviously, the anarchists don't take their demotions lying down. In Longines archival Worker Booklets (Serie E, 122.3), the historian Eitel finds "insubordination" and "unauthorized absences" that just happen to match the times and dates of famous anarchist speeches delivered locally by figures such as Pyotr Kropotkin. Essentially laborers walk off the job and subordinate the Longines work to their political action.

In July 1873, before the spying, Longines installs a steam engine whose first operation each morning is the official starting signal to everyone in the various departments to begin work. Two anarchist watch-assemblers refuse to submit to this new, highly impersonal factory-time: one is Samuel Schneider and another is Adolphe Herter, an engraver. They simply refuse to step into the factory and to their benches at the precise moment the steam engine powers up and sets a bell ringing – the penalty for which is not only a threatened dismissal but a blacklisting from all watchmaking firms in the region. By 1879, the penalties for all wasting of time are formalized in new rules promulgated by Francillon, per the American model of efficiency. Ceasing work before a machine has been powered off can earn a bench laborer a 50-centime fine (possibly a quarter of a worker's daily wage).

It is perhaps unsurprising, then, that Ernest Francillon, growing daily more draconian in the archive, moves almost seamlessly from industrial command into formal political office, as if the factory and the state were adjacent rooms in the same house:

Francillon is a deputy in the Grand Council of the Canton of Bern from 1878 to 1882, a member of the Swiss National Council from 1881 to 1890 and serves meanwhile as first president of the Conseil General (municipal legislature) of St. Imier in 1887.

He steals American tech and finds himself installed as head of the town.

And so we can say with not inconsiderable certainty: The Swiss-Longines spying effort is not just about quashing America's time-telling industry. It's about undercutting a workers' right to protest at the very height of the great 19th-century social struggle. Francillon successfully bets that the Swiss can win twice-over with American methods and machines: economically and politically (by keeping the working man down – and less needed and less well-remunerated consequentially).

There is a single man at Longines who can confirm all this – should he want to – Daniel Hug, the "Head of Brand Heritage."

He does not want to.

First, I mention to him the intellectual property heist.

Hug replies: "I just read your lines…I think you present them in a rather speculative way, which is for many reasons questionable." He tries to convince me that Jacques David, the Longines technical director sent to report on the American companies, is somehow justified in doing so because the whole world is attending the Exhibition in Philadelphia in 1876 – it wasn't like he was sneaking into the US at a time when no one else was visiting. He adds that "Jacques David himself started to push mechanization of production from 1867 onwards," as if the *planned* Longines' factory overhaul somehow means they had no need to copy American machine design in 1876, when US machines are indisputably better made and more efficient. And also when David steals into the Waltham factory in disguise.

But the exchange heats up further.

I mention Ambrose Webster, the former superintendent of the Waltham factory whom Gribi and David use ostensibly as their inside man and source of proprietary info they'd never otherwise have been able to acquire.

Hug does not reply for two months, and when he does, it's with an attached Word document of 2,000 words. Hug: "I think some of the terms you used are quite strong and refer to illegal and disloyal practices aimed at obtaining strategic information without authorization through the violation of secrets. We contest this view."

Hug adds: "In 1876 there was not yet a register of intellectual property, at least not in the US and not in Switzerland. This means an invention like a tooling machine or a new caliber of watch could not yet be protected by a patent."

Gabe Oppenheim adds: In 1876 – the very year Gribi and David spy on the Waltham watchmakers – sees a major race between two great minds *to register patents* for the telephone – Alexander Graham Bell versus Elisha Gray. This is front-page news. On Valentine's Day, both men's lawyers file paperwork indicating the near completion of the telephone: Bell files a patent application, while Gray files a "caveat" or a notice of an intent to complete an invention within a year.

The records show that Bell's paperwork is the fifth batch received that day and Gray's the 39th. Graham Bell therefore gets both the glory *and the patent.*

Which is to say, by 1876, America very much does protect intellectual property via legal stricture. Via *patent.*

Hell, Waltham mechanical superintendent Charles Vander Woerd accumulates in this period *27 patents* in the field of horology.

Oh, and the US patent system dates to the 1790 Patent Act.

Probably, I should give up – but I ask Hug about the anarchism as it relates to Francillon and David's need to take from the Americans the modern factory, with all its modes and mechanisms of worker suppression, subordination of the individual to the assembly line. Hug:

> *I don't know what you think we are doing in St-Imier. The brand heritage team has experts in history (one of them was a student of Donzé), in economic history, in communication, in economics, in the science of archives. We are fully aware of our history, and have an archive that goes back to 1867, in some cases earlier, with thousands of documents and archive entries and registers. But what bothers me is the very negative framing you are choosing. Besides, the story around Jacques David and his trip to the US has nothing to do with anarchism. I don't see any relation.*

The connection: the technologies for assembly lines Francillon and David stole via spying were precisely the innovations used to undercut the substantial number of founding members of the Anarchist movement who worked at Longines. To keep them down.

But then, in his defense, I am indeed telling a story where what is done by the Swiss – the rip-off – is then done back to them a century later. Our story takes us to San Francisco – gusty, frigid, a cheesy hotel with an aviation-themed lobby bar.

We wait for the man with the suitcase.

7

Hong Kong by Way of the Golden Gate, 1966

You've been here before – San Francisco. Eight years earlier, 1968, the height of all that hippie nonsense. You stared down sloping Powell Street from a perch just north of the intersection with Sutter – the movement all around you, undulating upward – a rising wave of taxies, trolleys, humans – disorienting and dense.

It was like someone were picking up the block and flapping it like a carpet to rid it of dust in a series of sine waves.

Those cable cars, one after another, pulled toward you by unrevealed, underground rope, seemed especially nightmarish – a ride you'd see in an amusement park dream from which you couldn't wake. All these humans being tugged toward grinding jobs they had to daily perform, lest they lose food and shelter – a "Twilight Zone" conveyance.

Yet as you descended the steep block, workmanlike and ordinary, you found you preferred it almost immediately to the highfalutin Nob Hill to the north and airy, manicured Union Square to the south. Here were institutions for the working person – the San Francisco of industry and elbow grease.

On the right, Grace Ball Secretarial College – not philosophizing Stanford or Cal — a school aiming to gain for its enrollees actual employment in the immediate future. Owl Drugstore on the ground floor of the college – the smell of its lunch counter burgers wafting out to you, the veins of savory fat sizzling against the stove-top.

Across the way, the Sir Francis Drake Hotel with its Starlite Roof, a 21st-floor nightclub with views of the entire city. A hotel that hid liquor, conveyed by rum-runners into Moss Beach, for its Prohibition-era guests in a secret room above its lobby – a whole mezzanine level dedicated to hooch accessible only by one of the four lifts and even then, only with a certain code.

One block south, on the Western side of the street, you found the Manx, built as part of the city's post-Earthquake reconstruction in 1906 and 1907, the Manx actually *ran* a speakeasy during Prohibition. And a decade later, it was the temporary home of the death-defying laborers who'd built the Golden Gate Bridge.

This block was layered with stories like gum on a sidewalk.

And finally, at the very end, where Powell met up with Market – Zukor's – middle class dresses for middle class women. Maybe upper-middle class. But in any event, more ordinary women than you might find in the city's trendier boutiques.

Yeah, you liked this block fine in 1968 – it had seemed ordinary enough for you — not that you knew then why its plainness might be essential just 8 years later. Not that you had any inkling you'd ever make a conscious effort to return.

Things had gotten tough. You'd taken over a jewelry and watch shop on the outskirts of town from an uncle – well, he wasn't *really* an uncle, just a family friend whose wealth always seemed of shady provenance – only to find profit a hard thing to come by, which was maybe the reason your uncle had been looking to off-load it.

People were always coming in asking for Swiss watches – Omega, Rolex, Longines – without realizing the capital a guy needed just to import them. The wholesale price was killer. Add in a $10.75 tariff per watch and a combined city-state tax of 3.5% on each sale – plus the rent just to keep the shop open and the electricity to keep the lights on – and the margins were impossibly thin to non-existent.

You'd already fired everyone but yourself.

And it wasn't even a level playing ground – you'd heard of other jewelers selling watches assembled in Hong Kong with simple Swiss movements and passing them off as the real McCoy – you'd heard of import deals where the importer was no more than a man with a suitcase – a hotel jobber, whose supply was all a Samsonite could fit and no more and who accepted in return for his wares only cash – on the spot. Someone who'd gone through customs declaring absolutely nothing and gotten away with it.

And then you saw the ad – first in the Chronicle and then in the Examiner on the same day, March 25, 1976 – just below an ad for abortions and above one for divorce without an attorney:

ATTENTION WATCH Importers Do you want to make your purchases cheaper? We can supply you Hong Kong & Swiss watches at lowest prices. Please contact Mr. Mahtani, Chanceller Hotel 362-2004 room 204

You knew it sounded sketchy. An endeavor only for the desperate. But what sort of person was scanning these back-page ads besides the despairing – those pressed into a corner already? You had to be honest: Once you'd started eyeing those ads, you'd already succumbed – you were the guy in their crosshairs with little left to lose.

So you'd called and set up a time and wound up back here – on Powell Street, staring down at all the establishments – secretary schools and former speakeasies, pharmacies and dress shops. And in the middle of it all, yet another almost entirely nondescript, unglamorous hotel, chosen for that reason – just as the second floor was selected so a potential customer could slip into the stairwell and avoid the elevators:

The Chancellor (that it was misspelled in the ad didn't change your mind; you were already all-in). True, the hotel's Clipper Ship lounge – decorated to resemble the interior of the earliest Pan-Am flights – was something of a magnet – a watering hole you might slip into if you were looking to pick someone up. But if anything, that bustle only served to steer attention away from the lobby, away from anyone coming to the hotel to do a different kind of business. Let them laugh it up in the Clipper Ship – it only left the halls of the place quieter and more anonymous.

You'd taken out of the bank your last decent wad of cash. Unless the guy pulled a gun and simply robbed you, you were gonna hand over that money in return for his hybrid goods whatever their exact nature (probably they were pin-lever watches assembled in Hong Kong with parts imported from Switzerland).

How would you sell them?

That depended, you supposed – whatever name was printed on the dial – you'd go by that.

* * *

The places are real. The San Francisco ad is real. The suitcase trade is real. Only our antihero is composite – and only mildly so. Hong Kong does not seize the watch trade with a drumroll or a master plan. It edges in through hotel stairwells, classified ads, cash deals, and men carrying inventory in luggage. That is how one of the great parallel capitals of Swiss watchmaking first comes into view.

And the Swiss help midwife the arrangement by shipping cheap movements eastward into Hong Kong. Only later do they grasp that they have been feeding hands that will bite them.

After years of protectionist railing against knockdown production, they have let their guard down – and almost definitely because they assume the Chinese incapable of ever reaching their industrial level in terms of quality or quantity. Consider what the president of the Swiss Watchmaking Federation found reassuring enough to file away in 1958 – a report from his envoy (or "spy," if one wants to be less generous), dispatched to a Chinese watch factory to assess just how likely the Chinese were to threaten the dominance of Helvetic horology:

"Seeing the semi-artisanal character of this factory where manual work plays an important role, one wonders whether the company will really manage to make the big leap forward next year."[103]

The envoy left the factory floor satisfied. The Swiss filed the report and moved on.

* * *

77 Queen's Road Central, the Central District of Hong Kong, the 1950s.

The colony now counts 2.6 million residents, nearly two thirds of whom are refugees of the Chinese Communist Revolution.

But Hong Kong, long the great importer-exporter at the tip of China — the mainland's proxy entrepôt — Asia's great global exchange — has a serious problem: A 1950 UN Embargo on trade of strategic commodities with Red China means Hong Kong can no longer do business with one side of its vast marketplace. It can't acquire Chinese goods nor can it hawk international wares on the mainland.

And so Hong Kong is forced to begin manufacturing on its own: It weaves textiles and garments, produces small plastics (including, rather famously at the time, plastic flowers) and falls back on an old trade at the harbor — shipbuilding.

Industry grows rapidly: In 1955, 91 percent of manufacturing establishments [employ] fewer than one hundred workers.[104] But by 1959, 75% of Hong Kong's exports are goods manufactured within the colony.[105] And 18 percent of those exports are bound for the United States.

A Time Magazine reporter, descending by plane into Hong Kong at the end of the decade, will thusly describe it from the sky:

> *The cluster of green, sugarloaf islands huddles close to the China coast. As the jet airliner glides in, sunlight reflects from the rippled sea, the brown batwing sails of Chinese junks turn in the wind. The travelers look down on rocky hills with terraced fields, deeply indented coves alive with sampans, a wide harbor carrying a honking traffic of freighters, tugs, barges and ferries.*
>
> *The jet swoops past the great city rising from the water's edge toward the towering Peak—shipyards, smoking factories, villas drowned in gardens, balconied tenements, squatters' huts clinging to bare rock, bright new skyscrapers still wrapped in bamboo scaffolding.*[106]

Queen's Road is itself a testament to the ever-changing nature of Hong Kong. In the 1840s, the street represented the northern shoreline of Hong Kong Island. Waves lapped the thoroughfare. But 100 years and several land reclamation projects later, Queen's Road is now two blocks — perhaps 200 yards — south of Victoria Harbor.

Still, the air can carry from the north the saltiness of the sea, the choking diesel fumes of ships — and from the west- from a stretch of Des Voeux Road West known as Dried Seafood Street — the overwhelming brininess of seafood arrayed in the sun.

It is amidst this bustling commerce, within range of smog, of scents marine and mineralic, that the watch shop at 77 Queen's Road operates. It is called "Shui Hwa" - which literally means "Switzerland Chinese" — and it's run by Hee-Kang Lam, who previously managed another watch vendor's shop in mainland Canton (Guangdong, as it's called today).

Lam's offerings are substantial and enticing and entirely legitimate.

How legitimate? In 1952, the Duke of Kent stops by, browses for forty minutes, then walks out with several Omega watches on his arm. Nine years later, Tetsu Katayama, a former prime minister of Japan, visits the Kowloon branch. Shui Hwa even acquires the exclusive regional rights to distribute high horology brands Piaget, Corum and Baume & Mercier, as the ad below suggests:

Meanwhile, as the legit watch merchants prosper, other immigrants to Hong Kong — the Mahtanis seem to be from a region of the Indian Subcontinent — discover an underbelly whose operations run in parallel to the legitimate watch-dealing of Mr. Lam at Shui Hwa – a line of commerce whose grey market-nature makes it the perfect work for a minority seeking employment in a territory it scarcely understands.

Essentially, the job requires men to stuff their suitcases with grey market watches in Hong Kong and show them off in hotel rooms in America to prospective buyers — thus garnering orders for the goods back home. (They can also take measurements for suits that they then have made in Hong Kong.)

But what exactly is inside these watches advertised from luggage — who has made the mechanisms — what is their provenance?

* * *

For a spring-barrel watch – that is a mechanical wristwatch in the 1900s – to unload that torqued-curve potential energy in a measured manner, its release must be controlled by an escapement – something that will hold up an energy discharge inclined by nature to speed ahead, to occur all at once, in a sense.

Every mechanical watch needs an escapement – a device that metes out the energy of a wound spring in measured beats instead of letting it uncoil all at once. Thomas Mudge invented the lever escapement around 1755, and it evolved into two versions: the jeweled lever, which uses tiny gemstones to reduce friction and keep time precisely, and the pin-lever, which replaces those jewels with cheap metal pins. The pin-lever is far less accurate and wears out faster – but it costs a fraction of the price to produce. Georges Frederic Roskopf built his "Proletarian watch" around the pin-lever, and the mechanism took his name. Such pieces became known as Roskopf watches — that is, when their vendors were being honest about their guts.

By 1970, the Swiss produce about equal numbers of jeweled-lever and pin-lever fully assembled watches – 38.3 million units for jeweled-lever and 32.7 for pin-lever. But open the ledger and the disparity is staggering: the jeweled-lever watches are worth $440 million (in 1970!), the pin-lever watches just $85 million.[107]

You can sense already how someone might take advantage of this value disparity to sell a watch with one escapement while pretending it uses another.

***

The Bejeweled Pallet:

An adroit, old Swiss watchmaker — achy and stiff, skin crinkled like a raisin — hunches over his or her wooden desk, a contorted body receded into itself like a flipped-over cockroach. The gangly and wizened tinkerer has but small hands free from the failing body to manipulate his or her tools. But that's all this watchmaker needs.

He or she holds the pallet fork over a plate heated from beneath by a small, table-top lamp burning alcohol. The metal softening, by controlled flame made malleable, the watchmaker applies viscous shellac to small dimples in the fork tines' liquefying tips. He then collects with brass tweezers the two red trapezoidal gems arrayed before him like tiny candies. These have been cultivated in a lab — they are the crystalline form of aluminum oxide adulterated with just enough chromium to render the microscopic structures vividly red. Rubies. Under a magnifying lens of some kind, the watchmaker carefully slides the synthetic rubies into the pooled shellac in the molten fork edges, the procedure requiring the utmost delicacy and precision — because all is asymmetry at this stage:

Each ruby has one uniquely angular face — the side, shaped differently by design, that will slide against the escape wheel's teeth, locking and unlocking against those lubed, scalloped surfaces, regulating the uncoiling power of the watch's gear train. Every millimeter matters here — the watchmaker must carefully alter and re-alter the rubies' placements, their depths, until he's found the exquisite points where they lock against and slide across and draw back in freefall from the teeth in optimal measures, such that the least amount of energy is lost to friction.

Now, the escapement is no longer free, the escapement has been confined — to a brushed metal case (no that matter that its name conjures the feeling of liberation on the last day of school). You admire this pallet fork set in position, this higher-class regulator of energy emancipation adorned with jewels like red jello, assembled by a fastidious watchmaker whose greatest successes will likely never accrue to his or her name — who has reconciled his or her entire professional existence to victories miniature and repetitive and anonymous. Devotion.

Is it any wonder the Swiss feel nothing for the many pin levers they produce?

* * *

The Swiss hit on an idea that seems, in the short term, clever: Why waste time building pin-lever watches fully at all? How much added money do they get from retailing a complete build with such a simple interior? Why not just export the cheap pin lever escapements uncased? Why not dump low-horology components in great numbers on that growing colony off of China that has suddenly become a manufacturer?

Re-enter Hong Kong.

By 1970, the Swiss are exporting nearly 85% of their uncased pin-lever movements to Hong Kong — about 22 million per year — *enough to case and ship 22 million watches back out to global low-end markets.*

But the Swiss don't comprehend that figure's implications — because no matter how fast Hong Kong is growing, there is no way it can produce that many *legitimate* watches. Those excess parts floating around the colony will provide a nice supply for the colony's franken-watch assemblers — those who throw together parts and likely call the pieces by a name they haven't earned. In fact, escapements are one of the last elements those Hong Kongers require to assemble franken-pieces. The Swiss already source many of their cases and dials from Hong Kong, so locals have sufficient expertise to produce those components for unauthorized watches. Now, without intending to, the Swiss have given them the watch's guts — the final puzzle piece.[108] It matters not that it can't be waxed rhapsodic about quite like the lever with jewels.

Thus is born the parallel trade to the aboveground Shui Hwa luxury watch distribution.

Some of the Hong Kong watches bear brands similar to Longines or Rolex but with a slight misspelling. Some are marketed as Swiss watches and others merely as containing Swiss movements. If you take the trademark infringement off the table – because many of these watches bear alternate brand names anyway – it's not even certain these watches fall entirely afoul of the law.

Of course, you'd need intrepid souls to sell them all over the world – a million tourists visit Hong Kong a year but business must be even bigger than that – recall that 22-million-unit capacity the Roskopf movements permit.

Reenter those immigrant families hawking watches out of suitcases — the hotel jobbers our noirish antihero returns to on Powell Street in San Fran in 1976 to meet up with at the start of the chapter. The ones who advertise in the back pages of the newspaper, darkly signaling to those in the know or those down on their luck goods on the cheap. You can be a seller, too!

And they do not just visit the US on rare occasion – this network crisscrosses the country, retailing from hotel rooms watches of mixed provenance (perhaps an expensive case paired with a cheap pin-lever movement or even a Soviet-made one) but also tailored suits and dresses – they invite men and women, via their back page ads, to hotel rooms to have measurements taken and fabrics selected, which choices are then relayed back to Hong Kong, where tailors cut the fabric and mail the finished products back to the US customers (all of it bypassing sales tax and duties).

Below, Mahtani ads for tailoring in Holiday Inn hotel rooms in Florida and New Mexico:

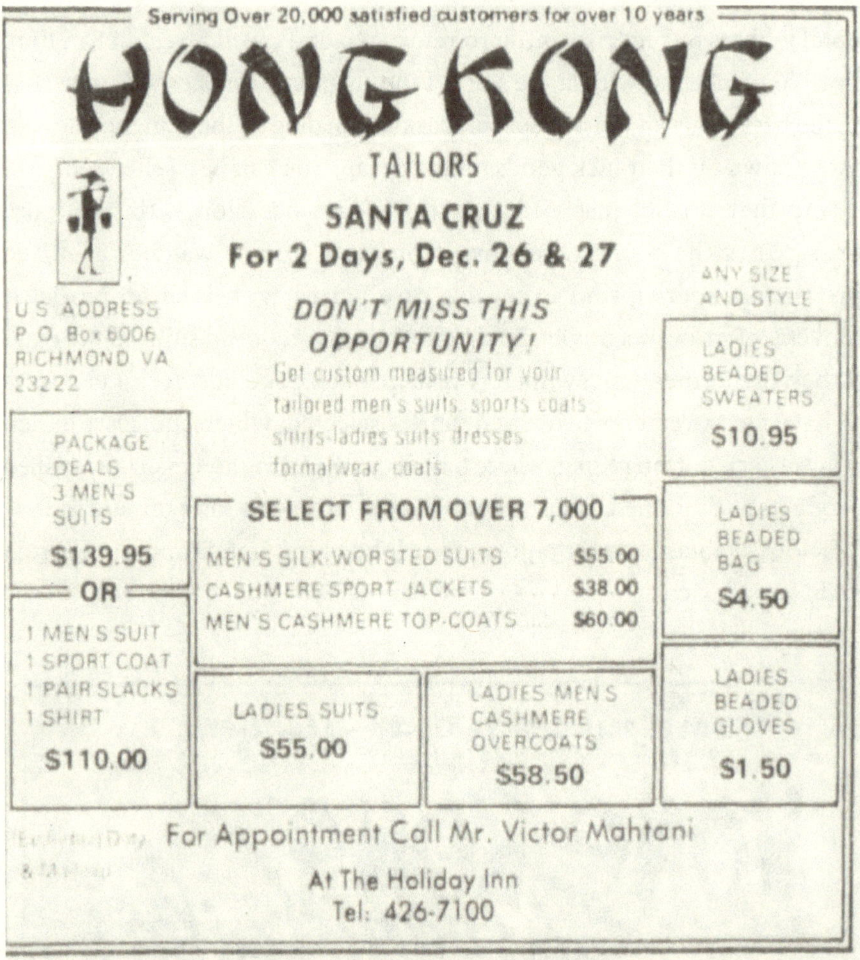

Of course, the Swiss are their own worst enemy in the development of Hong Kong as a source of muddled, mixed-part, pseudo-fraudulent inventory. They know full well they are providing more pin-lever movements than Hong Kong's above-board brands can ever need or want. The Swiss simply can't resist the good money they make off dumping such cheap movements – and they never for a moment suspect that the final product, its composition debased by that interior Roskopf, will ever steal away market share from their finely finished, jewel-pallet goods. And they are right – the Roskopfs won't – but the counterfeit trade they have seeded with such exportation – that will evolve, become further refined, grow into a serious threat indeed.

And it will be fostered in part by the man set up to protect the Swiss from competition. He is the president of the Swiss Watchmaking Federation – the Fédération Horlogère, the FH – and his name is Gerard Francis Bauer.

But on November 2, 1966, Bauer unintentionally does his own cause great harm. He signs an agreement with Sir Sik-nin Chau, a Hong Kong banking chairman and politician who formerly practiced medicine as an ophthalmologist and otolaryngologist.[109] The deal compels the Swiss to provide technical training to Hong Kong manufacturers so they can produce better quality dials, cases and bracelets for those Swiss companies that would like to outsource production to them.[110]

The Swiss make the Swiss watch Chinese.

Naturally, this offends a component of the cartel Bauer supposedly represents — the trade association of parts-makers – UBAH – one of three signatories to the 1928 Watchmaking Conventions. Their jobs are being outsourced to some small colony's unpedigreed manufacturers – who are going to be trained *further*, to replace them, for free, by the very Federation charged with UBAH members' protection (this was the point of the 1928 agreement – they were all supposed to protect each other).[111]

Eight days later, on Nov. 10, UBAH meets to review this tectonic shift in their industry — deciding to bash the agreement and the man who made it, Bauer, in an editorial to be published November 16th. Its headline: "A policy of the abandonment of the Swiss Watch."[112] UBAH makes clear its grave "concern and indignation."

The industry is now going far beyond the dumping of measly pin-levers into Hong Kong for nominal prices, which escapements find their way back to the US inside franken-watches retailed from suitcases.

And UBAH isn't alone in its dissent.

In fact, one particular company (Bourquard SA) makes its issues with this about-face so well known to Swiss officials that the complaint winds up reaching the desk of Hans Schaffner, who is a member of the seven-person Swiss Federal Council — the federation's executive branch. Essentially, one of seven leaders of the entire country.

This is how a Swiss government memo outlines Bourquard's full grievance:

> *"Firstly, the 'massive' increase in watch case imports from Hong Kong, which it attributes to the agreement made last autumn between the FH and the 'Federation of Hong Kong Industries'; secondly, the fact that these cases bear the 'swiss made' (sic) imprint; and thirdly, the imitation of their own case models by Hong Kong competitors."*[113]

Hans Schaffner, one of the seven heads of the Swiss Federal government, does not know what to do with this complaint (or he simply doesn't care to deal with it himself) — so he forwards it to his trade guru Raymond Probst.

Everyone can basically agree: There are three forces tearing the Swiss industry apart — the increased competition from upstart watchmaking nations such as Japan, the counterfeiting and franken-watching fabrication of Hong Kong and the need somehow to retain the prestige the Swiss have obtained (or stolen or both) while still cutting costs and becoming leaner and more vertically-integrated (and likely further outsourced).

Gerard Bauer is hearing crazy stories from the field as a result of watch salesmen feeling the pinch from these forces: In Japan, Swiss industry representatives are leaking to gray-market vendors Swiss sample collection watches – those models, entirely legitimate, that the brands produce for marketing purposes alone.[114]

Bauer must take action, but he knows he's penned in — the brands will always seek the cheapest way to produce their goods, even if the greater Swiss brand might suffer if customers wise up to the shenanigans. Neither can the Federation compel all these small watch case makers to combine forces (some of them would render the industry more efficient merely by disappearing). Counterfeiting and seeking new revenues in new places — these are the only areas Bauer and the FH are equipped to handle.

In 1971, he sends a roving delegate to Asia – one Dominique Thomas. The appointment itself tells you how outmatched the Swiss are: Thomas takes the post in mid-February and by the end of June must visit Bangkok (thrice), Vientiane, Hanoi, Singapore (twice), Tokyo, Seoul, Jakarta (twice), Manila, New Delhi, Bombay, Karachi, Kuwait, Bahrain and Beirut – in roughly that order. One man, seventeen cities, a continent and a half of trade issues to deal with, markets to open up, sales to spur. Though he is based in Hong Kong, he is hard-pressed to focus on that island.

And so Bauer refines Thomas' role, so that he — on behalf of the Federation — can concentrate on the lone issue perhaps resolvable:

Counterfeiting. In Hong Kong.

Step one: On September 2, 1973, at 10 a.m. a full Swiss delegation to Hong Kong files into the office of E.P. Ho, director of the Hong Kong Department of Commerce and Industry to finally have it out — the blow-out argument over Hong Kong fakery.

Leading the Swiss side is Raymond Probst, the Roving Ambassador for Trade Negotiations - a man whose father worked in the watch industry and who will, in just six years, be the Swiss Ambassador to America, the go-between during the Iran Hostage Crisis. That's who's underlining Switzerland's plight on this morning in this harbor at the end of the world.[115]

(Bauer is likely a more powerful figure in Switzerland than Probst, actually, but only the latter has a governmental title - which means something here.)

Earlier in the year, Probst has liaised with two Federation officials visiting his office to prep for the confrontation. One, Max Hool, is responsible for overseeing all counterfeit issues worldwide for the Federation - and the other is the roving Asian delegate Monsieur Thomas. Probst puts the question hanging over a possible diplomatic showdown rather bluntly:

> *"We need to know whether there has really been an improvement—as the latest reports suggest and our actions in Moscow give us reason to hope—in the efforts of the government of Hong Kong and the courts to combat watch fraud and other types of fraud more effectively,* despite the excessive British fairness that the Chinese in Hong Kong abuse and the danger of seeing job opportunities disappear, *which are certainly illegal but nevertheless 'welcome' for various reasons."*[116]

The encounter runs two hours and ten minutes. Various others from Hong Kong attend, including the superintendent of the Commercial Crimes Office.

"We would like to thank the Hong Kong authorities," Probst begins, deferentially and with a likely lack of enthusiasm, though the transcript doesn't admit as much. But he continues:

"The Swiss government cannot stand idly by and watch such a development, which affects one of our main industries and damages the reputation of our export-oriented economy."

Switzerland has long given Hong Kong advantages in its tariff regime it never earned in the first place.

"Whether this preference can be maintained and whether it will be possible to grant the second preference level next spring in the watch chapter also appears questionable, at least if there is no significant improvement with regard to counterfeiting. We would be very concerned in Bern."

The threat is not veiled. But Ambassador Probst has not come to this meeting either empty-handed or without a solution. Last year, Probst reminds his hosts, we presented "as a new weapon in the joint fight against counterfeiting, the draft bilateral agreement on the protection of industrial property. Last year, Hong Kong believed that it could not agree to this idea" – this occurred in an exchange of official cables in July of 1972.

"However, with its growing export industry, Hong Kong should now also have a growing interest in protecting its own designation of origin abroad on a reciprocal basis."

There it is – as clear as day – Hong Kong and Switzerland can blaze a path in this new interconnected world: they can protect each other's intellectual property for mutual benefit. Mind you, the Swiss did not do this with Waltham and generally, the entire history of horology is one of theft. Also, the Swiss claim this agreement is similar to ones they have with other European countries – so it would not be entirely new. But signing it in the midst of raging mimicry and intellectual property theft to put out the fire? Novel. Powerful. All Commerce Director Ho has to do is sign on the dotted line.

He won't. "Hong Kong," he says, rather evasively, almost as if he's paraphrasing the last line of "Chinatown," "does not attach undue importance to the issue of bilateral balance of trade."

At this point, the superintendent for the Commercial Crimes Office – Mr.

J.H. Grieve — speaks up. He tried using Interpol info to track down Hong Kong watches that had made their way to India without luck. And he knows that he has busted only small market-dealers – he has been searching for Big Fish – honestly, he has. But it would help if he had more info, if you really want Mr. Big caught.

Commerce Director Ho reenters the conversation: "As far as watch counterfeiting is concerned, Switzerland can continue to count on the help of the Hong Kong authorities in combating these abuses. [And I] doubt that other countries will be as accommodating in this regard—"

Immediately, Ambassador Probst protests – lots of countries help us a whole lot more – including those that have signed treaties protecting industrial/intellectual property, he retorts.

Ahh, but we won't be signing any bilateral treaties, Commerce Director Ho says again. "As a trading center and free port, it would be contrary to Hong Kong's liberal interests to enter into such a commitment."

As a free port: in other words, we'll be as lawless here as we wanna be.

And so after several similar exchanges the meeting ends with little progress being made – except, well, for this one suggestion, which seems small now but won't in future decades:

Yes, it's true, the Swiss are now in touch often with people on the ground in Hong Kong. The Watchmakers also have a roving Asian officer, indeed. But says, JD McGregor, assistant director of commerce of Hong Kong, you're not taking all the actions you could be.

"I've recommended on several occasions to representatives of the Fédération Horlogère," that is, the Watchmaking Federation, he says, "that they appoint a permanent agent for Hong Kong whose task would be to take care of the protection of industrial property in the watch sector on a uniform basis."

No moving between cities – a man dedicated to Hong Kong alone (and they are all men in this story, as it so happens, for better or worse).

And this is exactly what happens – though it takes some time before its dividends are apparent.

* * *

The first local overseer posted to Hong Kong to monitor watch piracy (and hopefully intercede) takes his post in 1974 – it is the aforementioned envoy Dominique Thomas. Every lunar new year, Thomas hears dogs cry out and he is startled and confused – until he finally realizes it's because they are howling on the way to being slaughtered. He tells this story to a father-son duo with the surname of Dubois (he's actually the godfather to one of the Dubois boys).

The father, Charles Dubois, a debonair watchmaker who taught the trade in Lagos, Nigeria, and whose family later lived in Lebanon until it was forced to flee the civil war, will wind up taking over Thomas' post in the 1980s. Sort of. At first, the government does not believe Charles has what it takes to root out fraud. So Charles begins his work in Asia on a highly quixotic personal (capitalist) mission. He will identify fakes and their makers himself on behalf of the Swiss companies — and he founds a company to do so — "Selective Swiss Watch Services" — a name eventually changed to "Selective Trademark Union." Only once he proves a success does he earn the official Swiss government position.

(Charles has always been a bit of a freethinker; he once bought up many lacquered ceramic ducks in Thailand because he was convinced he could resell the tchotchkes for more back in Switzerland; one year, he worked every night writing a book about gold, which he eventually published in Switzerland.)

But he proves so successful Chanel asks that he monitor the area for fakes of its non-watch products. His other early clients are a proxy company called Business Control that handles ugly legal matters like counterfeits for Cartier (and the other brands that will form the conglomerate we today call Richemont) and Rolex.

The son, Thierry Dubois, will assume this position from his father. Which is why Thierry is both official Swiss patrolman and private investigator both.

Meanwhile, Hong Kong and China will undergo their own generational transformations:

Even in Thomas' time, Hong Kong is on the cusp of being entirely overshadowed and outdone in every respect as a counterfeiter of the finest watches by the area formerly known in English as "Canton" but now more accurately called "Guangdong."

The mainland is no longer the economic backwater, the agrarian commune, it was during Probst's contentious meeting with Commerce Director Ho. And the Dubois family will have to deal with the consequences of its rapid financial and technological development: the corresponding rise in the quality of the counterfeits produced by its artisans and traders.

The birth of the super-clone watch in a land overwhelmingly connected, complex, futuristic.

8

The Great Opening, Mainland China, 1979

As 1979 draws to a close, a spokesperson for the city of Shenzhen, the mainland municipality across the border from British Hong Kong, agrees to meet a foreign news syndicate reporter in a modest, low-slung train station that will be demolished and replaced in four years. [117] Presently, no building in the city exceeds three stories.[118]

The spokesperson, Peng Peng, has never conducted such a economic pitch before and brings an English translator. Technically, this place did not exist before this year. Physically, sure — unfarmed hills, sloped soil exposed to the sun, overspread with tropical thatch grass (slightly sweet to smell if you snap it in two, but prone to catching fire); stagnant lowland marshes, muddy puddles of centimeters-deep standing water, twice yearly planted with rice, sulfurous, sour, vegetal-scented from the stubbled residue of uncollected grains.

But in January, this county called Bao'an is reclassified as the city Shenzhen. In April, Deng Xiaoping proposes an economic pilot in Shenzhen — a "special export zone."

The rules won't be codified till the following summer. But that does not stop entrepreneurs across the border in Hong Kong from taking advantage of the cheap labor here already. Entrepreneurs ink 800 deals to produce small goods — jeans, *watches* — in Shenzhen before autumn.[119] An estimated 6,000 young Chinese high school graduates begin working in small factories.

And then there's this train station, Shumchun, per the older pronunciation of this area — the southern mainland terminus of the Kowloon-Canton Railway, the last stop before a 105-foot bridge over the river, on the other side of which is Hong Kong's Lo Wu Station.

For 30 years, no train has crossed the border in either direction. Passengers who were allowed transfer had to walk across the bridge between train cars. Now they don't. The whole line runs uninterrupted again.

Sure, in May, 100 peasants storm and besiege a government building on the Chinese side of that frontier, trapping communist officials inside for 10 hours, after a careful plot to sneak poor migrants into Hong Kong is foiled.[120] So badly do these villagers want permission to leave.

Peng Peng does not bring that up. The liaison is here to discuss the investment opportunities Shenzhen now offers the world.

"Shenzhen is twice as big as Hong Kong," Peng Peng tells the UPI stringer, per the translator. "It is quite close and transportation is easy...Land rent is very low and things come in and go out without duties. *The capitalists make great profit.*"

(The last line sounds derisive when I hear it in my head, but perhaps that's just me.)

"By next year, we will have more rights and more authority and this will be guaranteed by provincial authorities," Peng Peng says. In the 1950s, factories had to turn all their profits over to the government. Now, some factories remit only half. Forget turning over proceeds — Peng Peng floats potential multi-year abatements of all fees.

And yet, even while attempting to attract Americans, Peng Peng can't help but mention the potential negative effects of a market economy.

> *"Inevitably, we will be influenced by foreign culture. Some people have bourgeois ideas and will use the black market to get foreign currency to buy luxuries but most of our people are strong and can resist the negative things about the west."*[121]

Peng Peng treats the consumer with a condescension and disdain not dissimilar from that which Swiss watchmakers have occasionally shown aspiring watch buyers distant from the Alps. The irony: the very city Peng Peng is duty-bound to promote, veined with marshy rice paddies, ducks waddling through its long grass, will grow into one of the world's largest black markets — purveyor of the world's finest superclone watches.

***

The economic change is both incredibly swift yet almost painstakingly methodical. Like China is truly planning for the long haul while the West — specifically, the Swiss and US — only thinks it is.

April 1984. Americans watch President Ronald Reagan tour China — the first state visit since Nixon's the prior decade. Upon his arrival in Beijing, he's greeted with a 21-gun salute. At a banquet that night, he calls for "mutual respect and mutual benefit."

But at the very same moment, a 10-member trade delegation from Shenzhen featuring its deputy mayor is quietly at work in Houston. Nobody is following this contingent.

The Shenzhen delegation is looking to build a city back home whose wealth derives from its electronics manufacture and its offshore-drilling (hence its connection to firms in Houston). Already, Shell, Sun Oil and Texaco have outposts in Shenzhen.

This stop in Texas is partly sponsored by a local engineering firm that has gained the right to build a $40 million, 35-story mixed-use tower in Shenzhen — with 200,000 square-feet of office space, hotel space and a heliport.[122] The firm is called CRS Sirrine. They've built the Gulf Tower in Houston. For two years in the 1980s they will claim to be America's largest construction-engineering firm.

But "Chinese planners carefully shaped the project around a *technology transfer scheme.* When the building is done by late 1986, the Chinese who have helped with its construction will have gained invaluable knowledge in steel-frame and curtain wall commercial construction."[123]

The Shenzhen Development Center opens up a year later than the Texans promised, topping out at 542 feet (taller than originally planned also) in 1987. It is the first high-rise steel-frame building in the People's Republic of China and, by default, its tallest.

Also, by the moment of its opening: CRS Sirrine, the Houston engineering firm, is no longer a part of the project. Maybe the Chinese acquired all the knowledge they ever set upon obtaining from the American firm and never *really* intended to erect the Texan version of the tower.

Maybe the Chinese, after learning American methods, considered alternatives and found them superior. Either way, the Shenzhen development group has placed the project in the hands of a Japanese company with a very different design called Kumagai Gumi. The Japanese company has also opened an office in Shenzhen and also invites the Chinese to learn from its ways. The firm even brings 11 Chinese workers to Japan to learn a very specific welding technique.

Shenzhen is quietly becoming adept at nimbly pivoting from one plan to another. At taking advantage of its second option when the first falls through — it doesn't let disappointment over an initial failure cloud its sight of the larger picture.

It happens in the gas fields, the very reason the government chose Shenzhen for this economic opening in the first place — the great promise of its offshore petroleum reserves.

"Even the most conservative estimates are staggering," Newsweek magazine said in 1975 of this capacity. "Some Western oilmen and intelligence experts even equate Chinese reserves with those of the entire Middle East."[124]

The reality, as reported by oil execs on the ground a decade later: "Fundamentally, the first round has been disappointing and unsuccessful," says George V. Wood, general manager of British Petroleum's China operation, which has drilled 14 wells and has gained for that trouble *14 dry holes.*[125]

The Shenzhen officials simply embrace the other industry they have been cultivating — technology. This is most obvious in the sudden birth of tech firms whose names still resonate: ZTE in 1985, Huawei in 1987. Neither firm makes chips or consumer devices yet (even though "semiconductor" features in the Chinese version of ZTE's name). ZTE builds electronic *watches*, electronic keyboards, hair dryers for Hong Kong brands — it's a simple contact manufacturer, for now. And Huawei merely trades telecom equipment made in Hong Kong on the mainland.

This is partly how the watch takes hold in this soon-to-be horologically-focused region. As the result of a pivot from the planning for an oil fiefdom to electronics — the makers of which all seem somehow or another also to create watches or their components.

For instance: In 1982, CATIC - a major player in Chinese avionics - establishes a branch in Shenzhen. In 1987, CATIC starts a watch company - Feida. In 1993, now called Fiyta, the company is listed on the Shenzhen stock exchange.

But the innovators building skyscrapers in the marshes are about to be the beneficiaries of a technology transfer long-carried out on the federal level — and that's the other factor that turns this estuary dotted with the dung of water buffalo into a timekeeping capital:

A secret Beijing-led campaign to extract from the Swiss their specific chronological know-how centered on their embassy in Bern.

* * *

Decades before Deng Xiaoping declares Shenzhen open to industry, Chinese functionaries abroad decide:

We want Swiss watches ourselves - but we also want to be able to *make* Swiss watches.

The desires: As early as 1957, in the time of Mao, the Chinese buy 124 million francs' worth of Swiss watches - a number that only drops precipitously in the following years because of the economic crash caused by Mao's "Great Leap Forward."[126]

But in the economic rebound following the failed leap, from 1963-1966, of all Swiss exports to China, watches increase by value the most.[127]

The lust is further evident on the very wrists of Chinese diplomats. "The

Chinese know the best [Swiss] brands well," Swiss Ambassador to China Hans Keller observes in 1964, "since almost all higher officials here have Omega, Rolex, Longines, or other first-rate brands."[128]

Keller figures the Chinese diplomats are acquiring the timepieces in Hong Kong – these functionaries cannot all have passed through the Swiss embassy in Bern in their service careers, after all. Meanwhile, the administrative functionaries and staffers who do serve permanently in Bern are known to wear Omega and Tissot.[129]

In 1963, the Swiss Ambassador to China tries to convince his compatriots in the business community to take mainland China seriously as a real market — arguing that while not every Chinese person might be able to afford a Swiss watch *now*, their families surely will be one day (there's amazing prescience and planning there, even if the advice isn't taken to heart).[130]

All of the above attests to a distinct Chinese enthusiasm for horology that long precedes the opening up of the marketplace to free enterprise in Shenzhen.

But the Chinese desire is never the simple consumerism Peng Peng so harshly derides in her train station interview announcing Shenzhen open to world business. They always want to be able to *recreate* these timepieces.

Evidence of this long-term plan to acquire the *means of* production is evident as early as first decade following the Revolution.

The Swiss consul of Shanghai, visiting a new watch factory in the city in 1958, is shocked by the obvious presence of tooling devices produced by Swiss companies Hauser, Petermann, Bergeon, Lambert, Tesa, and Tripet.[131] These cannot have been authorized purchases — they have reached Shanghai likely via some third party. By 1964, 20 percent of watchmaking machines in Shanghai have been illegally procured from Switzerland.[132]

And the Chinese take measures to keep this build-up secret. In 1961, China bans Swiss citizens from its watchmaking facilities.[133]

Even the Swiss Ambassador to China, Hans Keller, is only allowed entry into a watch factory after the government has formally ordered the site to give Keller the run-around – the tour skips entire sections filled with equipment he shouldn't see and keeps far enough away from machinery such that he can't discern serial numbers.

The Swiss sense the chicanery, even if they can't track the exact black market acquisitions. From June 1964 through February 1966, the Swiss diplomatic corps and the Fédération Horlogère consider putting on a show of fine watches and even some machinery in China. But China's very specific prerequisites for a show are nonstarters for the Swiss. The Chinese want the displays to include the individual components of watches. The goal is very obviously a close study of protected Swiss IP — an unauthorized transfer of technology.

The show never happens.

However, the leaders of Swiss watch brands – less aware of China's watch espionage operations – are very eager to enter the country once it begins to signal a future market liberalization. Or maybe they are blinded by the commercial opportunities a Chinese opening will yield. They have a "gold rush mentality," according to one Swiss official – and so they ignore all the warning signs as they attempt to insert themselves into a marketplace ancient and complex.[134] They ignore Peng Peng's derision of consumerism — the longstanding dedication in this region to *production* above all, as if there is a great moral divide between the two sides of the market.

Longines and Omega open watch service centers in Shanghai in 1980. The following year a three-floor Swiss watch retailer opens up. The last of these venues employs 70 Chinese eager to learn the trade.

But as soon as the Chinese workers at these ventures have achieved a certain level of competency, they are induced to leave the job by the government and go to work in a Chinese-owned watchmaking operation.

The Swiss even open a watch factory in China – only for the government to force the manufacturer to raise wages by 10 percent regularly until the business is no longer sustainable – but a good number of Chinese workers have been well schooled in Swiss horological production. They use my factory as an "educational workshop for China's watchmaking industry," the factory's general director decries.[135]

Trade delegate Raymond Probst has laid out as early as 1967 the way China will take Swiss business while *doing* Swiss business, writing: "The 'trend' of increasingly sourcing cases from abroad, which supposedly provides cheaper prices and shorter delivery times, is well known."[136]

Back in Switzerland, the FH decides that if it can define "Swiss-Made" somehow, and restrict usage of that name on the dial, it will give its pieces an edge that no amount of unauthorized technology transfer to the Chinese can erode.

Even if two watches from the two countries are functionally identical, the title alone will signify an otherwise unattainable quality.

Swiss-Made.

Only nobody can agree on a definition.

in the late 1950s, the FH enlisted a Swiss theologian, Professor Robert Martin-Achard, to make a ruling, Probst writes.

"He concluded that a watch 'whose movement has been assembled and regulated in Switzerland' may rightly be labeled 'swiss made'. (sic) Under this rule, a watch with a Swiss movement, fitted with a foreign dial and encased in a foreign case, would be allowed to carry the 'swiss made' mark."

Probst says the head of customs for Switzerland was willing to adopt that definition — as were courts in foreign lands — Belgium, France, the Netherlands. But UBAH — the union of parts-makers and a part of the cartel —dissented — because, once again, their constituency included the very producers of Swiss dials and Swiss cases who'd be forced out of business.

But the FH is willing to give it another go, Probst says — and this time, they've solicited a proper definition of "Swiss-Made" from the watch industry's go to legal big shot, an IP expert named Jacques Cornu — according to whom "the decisive requirement must remain that the bridge and mainplate of a "swiss made" watch are Swiss.

Now, if that seems oddly specific, it really is: the bridge and mainplate are elements of a watch mechanism that essentially hold the entire thing together — the bracing for those elements that spin and whir and click. Under this definition, were it adopted, the gears and springs and jewels of a movement could be sourced from Hong Kong (or Italy or Thailand) — let alone the encasing structure or the hands or dial.

The idea is such a baldfaced attempt at allowing the maximum amount of outsourcing possible while retaining the notion of a homegrown product. The movement's scaffolding is almost never the most intrinsically interesting element of a watch (one exception that comes to mind is the Girard-Perregaux series of watches that feature three golden bridges, but that's the exception that proves the rule).

Probst concludes his letter to the head of state: "The overall situation remains unclear and unsatisfactory… In the long run, it will likely be unavoidable for authorities to participate in the search for a widely recognized regulation of 'swiss made' (sic) in the watch sector."

Indeed, the entire kerfuffle does lead to the adoption and acceptance of a final rule in 1971: 50 percent of the value of a movement's workings must be Swiss, and the movement must be assembled and inspected on Swiss soil. That's supposed to guarantee that local hands will always touch the mechanism.

You can't roll back globalism with a dictionary.

* * *

In 1981, China produces nearly 30 million watches, 95 percent of which are low quality and sold on the domestic market.[137]

By the 2000s, the Chinese are producing *hundreds of millions of watches per year* – the vast majority under the brand names of foreign companies, as Original Equipment Manufacturers (Private Label suppliers) – watches bound for export to all corners of the Earth.

In fact, when the very man who organized the first Watch and Clock Fair in Shenzhen in 1988 – Shunhua Zhu – is asked in 2018 to name the single biggest Chinese watch producer, he can't even name one brand.

"It's difficult to say," he says, "because the companies that produce the most are OEM producers" for *other* companies.

But he does peg total Chinese production at the astounding figure of *1 billion timepieces per year.*

All the while, a rule is in effect that in essence legalizes Chinese production of Swiss watches, which outsourcing has become a highly attractive cost-cutting strategy to historic Alpine brands suffering in a globalized world: that 1971 rule requiring only 50 percent of a watch's movement by value be Swiss for a timepiece to bear those two magic words on its dial — "Swiss Made."

(In 2017, the rule is slightly amended, so that fully 60 percent of *the whole watch,* minus the bracelet, must be Swiss to carry the designation, but as the cost of labor can count toward the 60 percent, you can get to 60 simply by paying standard Swiss wages to a few folks who inspect the watch.)

And so proceeds the rebirth of the Swiss: their fanciest brands feel free to acquire bracelets, dials, cases and hands in the Pearl River Delta. Where concrete is crowding out mangroves, oysters. But the truly revolutionary aspect of this moment in time is Swiss latitude with its movements – the requisition of movements from China and Southeast Asia.

ETA (part of the Swatch Group conglomerate), Ronda and ISA Swiss – all movement firms – open factories in the region. Some Swiss firms even acknowledge Swiss movements are being copied by the Chinese now — to the degree that they try to sneak back into Switzerland Chinese clone movements themselves rather than buy originals.

Valanvron imports what appear on teardown to be base Seagull ST2130 kits and upgrades them in Switzerland. The Seagull ST2130 is a clone of an old Swiss movement, the ETA 2824-2, patented in 1982.

And then there is Inter T, which in the 2010s imports almost all of its movement components, likely from China – save for the most expensive parts, such as the hairspring, to reach the golden ratio of halvsies.[138]

The company doesn't pretend otherwise, in fact – it claims that, as ETA stopped allowing companies outside the Swatch Group (of which ETA is part) to use more legitimately Swiss mechanical ETA movements, Inter T had no choice but to rethink its supply chain.

* * *

What's the greater legacy of the Shenzhen Special Economic Zone, of the first opening of the country, in the province of Guangdong?

What happens to all the watchmakers and assemblers it creates – artisans in their own right many of them? What about those endowed with knowledge sneakily acquired at Swiss service centers?

What if I told you that watches themselves, no matter the large volumes shipped by China, will experience a long period of creative stagnation until the mid-2010s, at which point a curious thing occurs – little companies with tremendous ingenuity begin popping up on social media, tiny enterprises of a few people with clever ideas and strong aesthetics – and that consumers order their plucky timepieces sight unseen by the boatload?

And what if I told you those boats come and go from Hong Kong's ports – because the microbrand revival fully avails itself of the many artisans and craftspeople, the factories big and small, of Guangdong?

That the microbrand trend is really what Guangdong has been awaiting since it first became a watchmaking haven in the 1980s. The Pearl River artisans can now connect directly with their customers – can even solicit feedback on Reddit about potential stone or guilloché or coral dials?

It's a level of homey personalization and convivial exchange the artisans always desired and consumers realize they greatly appreciate.

Oh, and one more thing: All around these honest benchworkers and computer designers, crafts-people of the new Millennium, beside their numerous suppliers, exist other assemblers, who often tap into the very same supply chains. They also reach out to their customers on social media, solicit feedback, show off prototypes. In fact, there are on many fronts almost no real differences in their methodologies.

Except this latter group produces huge quantities of gorgeous Rolex Submariners, Patek Philippe Nautiluses and Audemars Piguet Royal Oaks.

They also have been lying in wait, have also been reawakened by a new internet-driven public consciousness of horology – and they have all the blueprints to mechanisms the Swiss eventually just left lying around, basically turned over.

That long con – they are the endpoint – assemblers of watches aspiring to be 1:1 clones of the best Swiss present-day offerings.

They present for your consumption the results of a 75-year espionage operation.

Guangdong: filled with youthful, legit artisans and equally energetic counterfeiters (arguably with even more watchmaking skill, given what it takes to clone a movement, with its precise tolerances, so perfectly).

Guangdong: A province we therefore must enter more fully.

9

Pearl River Delta, Dandong, Padua, 2015

But before we enter the factories, a necessary detour – because the moral question at the heart of counterfeiting is older than any of the technology that enables it.

A slight discursion into the history of a 17th-century French physician and his son.

Dr. Guy Patin, the father, hails from a family of prominent attorneys who argued cases before the high French courts called "*parlements.*" But Guy rejects his legacy and chooses instead the noble pursuit of medicine. He then engages in a great deal of "bloodletting and purging" as a physician – this being "the farce-comedy era in the medical history of France."[139]

It is unsanitary, unscientific, and unintentionally hilarious.

Dr. Patin's son Charles is a remarkable figure almost from birth, in 1633. He can read by age 3 and write by age 4, attends a Jesuit college as just a teen and defends his thesis in philosophy at 14. He then also becomes a doctor – the real-life Doogie Howser – but one with a wide variety of interests, not all of them entirely appropriate or legal, let alone medical.

At age 30, Dr. Charles Patin acquires the coin collection of the Count of Brienne and delves with his great intellect into the world of numismatics.

Also...book smuggling – a hobby he pursues alongside his father, a bibliophile who perhaps merely overreaches in the same way he simply cannot cease making incisions in the operating room.

Pere et fils begin importing into Paris banned and counterfeit books created in Holland – a crime for which, once both are caught, only the son Charles faces sentence – to a lifetime of forced labor – a penalty the justice system then considers only slightly less harsh than execution.

So Dr. Charles Patin flees, escaping before he can be imprisoned properly, and roams the continent for months, before settling in Padua (at a time, recall, before Italy is a unified nation-state).

Here in Padua, perhaps Charles atones for the medical mistakes of his father: His obvious excellence in that practice earns him the position of chief of the medical faculty at the local university. But as so many passionate collectors and counterfeiters, Dr. Charles Patin cannot keep away from his earlier habits.

In this case, Dr. Charles chooses numismatics instead of forgery. He pens scholarly texts on ancient coins.

(He also delves into the potential use of boggy peat as a fuel – writing one of the earliest modern treatises on the usage of biofuels. And he encourages his daughters to study and write academic papers, as well – a most progressive rearing.)

Dr. Charles dies in 1693 but leaves behind that signature achievement in scholarship, his book "Introduction to the History of Medals," published when he is only 32.

In it, he divides counterfeiting into two categories – those who intend to deceive others with their forged coins and those who understand the aesthetic, historic value of a coin they cannot possibly obtain and, yearning for it anyway, acquire a copy instead (specifically, he is referring to a collection of coins from all 12 Caesars – as one such ruler, Otho, also known as Othon, presided over the Roman empire for just three months – hardly enough time to mint sufficient coinage for collectors).

"The effigy of Othon, or that of another prince, is no less represented to us in lead or copper than in gold and silver," he argues. "The history is the same. The information is no different."

And more directly:

"We are content with modern coins in the absence of ancient ones…It is better to have a fake than to have none at all."

* * *

Patin's distinction – between the forger who deceives and the collector who simply wants what he cannot have – is precisely the fault line that runs through what happens next, in a Chinese province still locked down by Covid.

It's April 2021. Much of the world is edging out of Covid lockdowns with that peculiar, underwater feeling the pandemic left behind — as if all humanity has spent a year speaking through glass. The streets are repopulating, the air loosening, but everybody still seems half prepared for the next closure.

There is, however, a Chinese city still cut off entirely from the rest of the world by authorities, under strict quarantine, its businesses and schools shuttered, its residents cloistered in their homes – Ruili, in Western Yunnan province.

Unfortunately, for its denizens, Ruili lies on the border with Myanmar, and nationals from both states cross the border often as a matter of basic commerce and everyday life. Myanmar does not have Covid controlled on its side of the linc, and so the virus has spread northward, infecting the Chinese of Ruili.

China's Zero-Covid apparatus drops on Ruili with the choreography of a raid — except the uniforms are white protective suits instead of black armor. The health workers move like a tactical unit because, in a system built on visible control, outbreak response must look not just effective but overwhelming. Two local Communist officials are stripped of their posts as well, punishment offered up with the usual bureaucratic theater to prove that somebody, somewhere, is in command.

On April 8, China reports 79 symptomatic cases and another 44 asymptomatic ones (in the end, no one will die from this particular cluster).

And if you follow the news, if you are a citizen of the world, certainly if you are yourself a Chinese national, this is a development of which you are made aware by the media. The world is still quite cautious generally — fearful of explosive new pockets of evolved COVID variants. Recall that the pandemic is barely a year old. How scary it is that COVID is actively morphing to evade our best medicines, which we only just developed.

These are days when it seems it all could end at any given moment.

And so it does – but not for the citizens of some remote Chinese border town or as result of our contemporary plague.

In early April 2021, far from Ruili, another Chinese strike is unfolding — this one in the watchmaking sprawl of Guangdong — that provincial home to Shenzhen, where all Chinese mainland watchmaking first emerged, in a sense, 45 years earlier.

Public Security officers descend on a factory after a long surveillance and begin lifting workers and managers off the premises with the practiced speed of officers trained to leave no time for shredding, hiding, or warnings one to another of any kind.

But as the walls close in on these arrested line-workers, as their world contracts in a matter of minutes from factory floor to interrogation room, none of it becomes global spectacle. This raid is swift and invisible.

There are no breathless cable-news hits, no chyrons announcing the fall of a counterfeit empire. Unless you already haunt the darker corners of the watch world, the whole event passes almost noiselessly, like a safe being rolled out the back of a building at dawn.

It's only the greatest Chinese raid on a counterfeit watch manufacturer in the country's history – the takedown of an operation that over the course of six years developed and distributed the best fake Rolexes and Panerais ever made, the finest fakes ever to hit the market – an enterprise valued (maybe conservatively) by the state at 300 million renminbi – *or $42 million.*

On the day of the raid, authorities confiscate a staggering 100,000 watch parts, in addition to thousands of completed timepieces.

But how did any of it happen – how were the principals able to create such faithful clones? And if they were so exceedingly clever, how did they ever even get caught?

And what of their international distributor, the man who funneled their work straight into the heart of the European continent – who in 2024 will be ordered to pay a fine of $220 million, based on the value of the watches he has sold had they been *real*?

Yes, that's right – more than a fifth of a *billion* dollars.

How does the Special Economic Zone of the Pearl River Delta become the Super- Clone Cloister of the World?

* * *

It begins, as so many stories in this book do, with a young man moving toward the seam in the market. He is a 25-year-old Frenchman from Nice, a onetime pizza-delivery driver, and he believes a better life is waiting for him in Guangdong, that sprawling province where Chinese watchmaking has learned to hide its brightest and darkest talents side by side.

Raised by a single mother, Julien Vincent seems to have learned early that legitimacy is something other people get handed. He counterfeits Disneyland passes at 12 — already drawn to the little thrill of making an unauthorized object pass through an authorized gate.

He drops out of middle school before obtaining his diploma, then appears in court with metronomic regularity: an annual court appearance between the ages of 16 and 21 for driving without a license, robbery with violence, repeated death threats with a knife, and drug possession.[140]

And then he commits his first crime involving fake Asian wares – he opens up in the early-2010s a Facebook page called "Mode et Bijoux Paris" to sell counterfeit Chinese clothing. But this is small game.

Now, Julien Vincent flies to Guangzhou, Guangdong – formerly known as "Canton City," this is the province's most advanced area before Shenzhen becomes an actual municipality. He is 20 years old, and he has but 5,000 Euros in his pocket – seed money. At first, he simply buys even higher quality knock-off leather bags and ships them back to his mother for resale in France (she ends up with 15-month suspended prison sentence). Then he spends Lunar New Year 2016 in Thailand and, entranced by the place, decides to relocate there (also, it's a great gateway to Asia for a criminal-minded trader).

Upon his return to Guangzhou, Vincent finally begins to meet the watch artisans. The Frenchman establishes a network – that is, he makes sufficient contact with counterfeiters that he feels he has a supplier locked in for a timepiece-distribution business.

He makes a home for himself in Thailand – either in Phuket or Koh Samui, where he believes himself safe from the law. He calls the Asian element of his nascent operation: "Prestige Replica."

So far this would be rather ordinary fraudulence – unexceptional – a Westerner looking to make use of the great Asian capacity for precise industrial horology.

But Vincent has not met up with anyone one might term ordinary but the very best China has to offer in this field of watch replication – with the Noob factory (alternatively, the Noob factory is formed – coalesces – only once Vincent has visited – but in either timeline, he has made damn fine connections).

Most Chinese counterfeiters to this point and since have operated in the traditional, horizontal horology ways of Guangdong – various suppliers provide dials, hands, cases. The counterfeit "factory" is really just a site of assemblage.

Noob is different. Its principals take over a real building – a proper facility with equipment and floor space, per later Chinese authorities reports. The scheme begins in 2015, and it begins with money: a man named Xu sinks a major investment into the fake watch trade, pulling several co-conspirators in after him. For three years, Zhuang runs procurement – but all he can secure are Asian copies of the Valjoux 7750 chronograph mechanism – a standard, wholly unimpressive engine for a mechanical watch – and not one found in a contemporary Rolex either. It has a horizontal clutch instead of a vertical one — meaning the stop-watch mechanism cannot engage with an immediate click. There's a hitch when you hit timing button before engagement.

Then Zhuang walks. He pulls out of the operation in 2019, seemingly of his own volition, and his replacement is Chen – a man who thinks out-of-the-box and is rewarded handsomely in return.

Chen does something none of his predecessors consider. Beginning perhaps as early as 2018, he does not simply rely on the best movement-maker in his own Guangdong province – he travels north – to a border region across a river from North Korea – to the old Peacock Watch Factory in Liaoning Province – that is, Chen travels to the city of Dandong.[141]

At this point, Peacock produces movements for aboveboard Chinese watches – for products you can legitimately buy across the country – and even in the west.

But Chen organizes a deal whereby the Peacock factory (perhaps in a backroom, perhaps more openly) actually crafts for him clone movements – in particular, a copy of the Rolex Caliber 4130 that the Swiss company uses in its highly sought-after Daytona Cosmograph (aka the "Paul Newman" watch). Which has a vertical clutch — its push buttons actuate immediate stop-watch action.

The movements are bound to be excellent. Peacock has every advanced tool and has been perfecting its craft for decades.

Here is a factory that knows something about copying – and about defying the powers above it. And about being punished — about the unjust rewards for compliance with the Chinese Communist Party.

In the late 1950s, Liaoning – the factory's original name, taken from its home province – was one of the first four watch facilities the state established. Its earliest mechanisms were frankly Soviet copies – the SL1 represented mimicry at a moment before the relationship between the two communist great countries fell apart. Under the brands Qianjin ("Forward"), Hongqi ("Red Flag"), and Kongque ("Peacock"), the factory turned out full-on watches – but it was Hongqi, produced from 1966 to 1978, for which it devised its first original movement, the SL2.

That was Sino ingenuity – via the Swiss: using Swiss machinery, Liaoning produced a movement that beat 18,000 times per hour, then updated it to the SL2A, basically the same engine but with an increased frequency of 21,000 BPH (the very same beat rate of the rock-solid Seiko NH35 movements found in many mainstream brands today).

And then the state stepped in and crushed it. In 1973, the Chinese government, for the sake of national standardization, forced Liaoning to quit using its own caliber and rely on a crap movement being deployed across the nation, produced by the Ministry of Light Industry. Worse, Liaoning was ordered to produce a million watches per year by reducing their manufacture and evaluation processes to nine steps from 22.

You are an engineer on the factory floor at Peacock in 1973. You live in a grey city on the border with North Korea. Your work is your escape from this dun, depressing frontier environment. You thrive in this metallic box that others would fine confining.

Imagine you are that Peacock engineer confronted with such massively unfair ordinance:

> *You have built up from nothing new movements. You have conceived of complex solutions to engineering problems no one else foresaw. And consumers have en masse flocked to your innovations. How rewarding! And now, standing on the noisy factory floor, surveying the glorious machinery at work carving up the components you envisioned and drew up, you hold in your hands the caliber — the clunky, inelegant movement — you must by arbitrary fiat insert into watches rather than your superior handiwork. The disappointment is a dark hole. You cannot reply that you've spent years of your one life to live on crafting watch movements with symphonic structure and harmony. The law has come down like a fist slammed on a table. The system has spoken.*

The result was predictable – and devastating. Hongqi was rated worst in quality across all of China. By 1978, the factory owners killed off the brand altogether and renamed the whole operation Peacock – "Kongque" (they actually switched off between using the English and Chinese terms; I have pieces under both names, though my Kongque dial boasts a cute little peacock emblem the English-language version lacks).

When Peacock reopened, it was still not quite allowed to pursue its own work. But its engineers figured out a very clever circumvention of national standards – they built a module on top of the mandated movement, converting it from a manual-wind mechanism into an automatic via an added-on bi-directional rotor. The jewel count jumped from 17 to 33 – jewels being just a means to an end in watchmaking, indicating the number of places where possible friction or resistance or scraping might be found were a jewel not used. The real innovation was that the wearer's wrist now powered the mainspring, rather than 30-something turns of the crown.

Yes, Rolex adopted this innovation a full 50 years earlier – but it did not have to contend with an authoritarian government forcing on it a crappy base. Peacock becomes the definition of free-thinking within a state that bans it: its automatic is built on top of government-mandated mediocrity.

A 33-jewel automatic watch beating at 3 Hz is the very best timepiece anyone can buy in the early 1980s – but it is competitive enough on features that the movement allows Peacock to export goods to Eastern Europe, among other poorer areas – for the first time.

And it is this facility in the mid-Aughts – some criminal element embedded within it, if not those at the factory's top – whom Chen, head of procurement at the counterfeit Noob factory, convinces to create for him the Rolex Daytona mechanism clones. With vertical clutches and instant actuation of the stopwatch function.

A tremendous procurement from a facility with a long history of impressive, aboveboard production. The facility that faced down the Communist state.

With the movements secured, the rest of the operation locks into place. Cai and Wu, two of Noob's remaining leaders, source dials, cases and bezels. Zhuang – the same man who'd run procurement before walking away – and a colleague named Wang are brought on to assemble the packaging and distribution teams. Both will depart by 2019, as though some instinct tells them the enterprise is edging too close to the sun. Will be raided.

The Noob exec suite is in place – it has suppliers of the highest order and a weird French exile living in Thailand guaranteeing them purchases of great quantity – it has a customer.

But Noob will come to regret sooner rather than later accepting that whale customer, one whose massive remittances are increasingly unnecessary in a world in which counterfeit buyers are well aware of the factory's quality.

Noob begins making watches in 2015 for the underground community of collectors – a loose affiliation of watch aficionados globally who often know more about parts, assemblage and watchmaking principles than the customers of the genuine models.

They congregate where the rest of us do – on whichever social media platform is prevalent that year – Snapchat, Reddit, Discord, WhatsApp, Telegram, Signal – even on some dedicated websites anyone can openly access, based in countries unreachable by Western law and disinclined to its adherence.

With the 2019 release of a Rolex Daytona with a perfectly-cloned (precisely, near-exactly reverse-engineered) Dandong 4130 movement, Noob obtains its greatest mass validation. Individual retail customers online come to a consensus – Noob's Rolex Daytona is of utmost quality, a greater achievement than has been made by any other replica watchmaker.

Noob works on Rolex watches most intensively – it is not expert in the production of other brands. And the French underworld trader in Thailand, Julien Vincent, also wants to sell Patek Philippe, Audemars Piguet, Chanel via his "Prestige Replica" operation.

Vincent finds Thai middlemen who will buy the other brands from other Chinese counterfeit operations, which I won't name here, in part because some are still active and this is not a prosecution. And my source on the inside of the bust has placed strict limits on what I can report.

Now that the Frenchman has a wide inventory acquired under his Prestige Replica brand, he sets about distributing it in his native France via a Telegram channel with myriad subscribers, overseen by his second-in-command: Florian Rebani – all this under the brand name "La Genèverie."

The business is brisk: Vincent over a half-decade releases between 12,000 and 15,000 watches onto the Francophone markets of Switzerland, Belgium and France.

And some are merely built off of Noob bases – in some he substitutes the Dandong movement for its legitimate Swiss counterpart; in others, he replaces the Noob Dial with one from an actual Rolex.

These are called Franken pieces, and while they're far more expensive to create than the Noob pieces unaltered, they can also be sold – *deceptively* – as legitimate. The prices range from 500 euros for the base fakes from Noob to 6,500 euros for the best frankens.

Tellingly, to many a Noob buyer, who has never acquired the watches through this French hustler but rather through agents called Trusted Dealers, who serve as intermediaries between Western buyers and Chinese factories, this last deceptive bit is the offensive part – the true crime (a feeling experienced only in retrospect – they're still unaware Vincent even exists).

The online counterfeit community, those who expound endlessly about the qualities of particular hairsprings and shock systems, appreciates the aesthetic and technical innovations of their clones. Like Dr. Charles Patin, they believe in acquiring for study and appreciation that which will otherwise remain unattainable to them.

Julien Vincent is all about making that bread – and he does so in such a public and brazen way he puts both his own operation and the Chinese factory that supplies it at serious risk.

He posts photos on social media showcasing his ostensibly glamorous lifestyle – his Lambo, his Porsche, his houses. The tattoos all over his body, including the ones on his knuckles (the left hand spells out "money" and right "cash").

He also launders money in a way that earns him attention, which seems highly counterproductive: He opens a shisha bar (which ends up in failure when his two partners start a rumor that Vincent is being tracked by Interpol so as to convince him to close down the place — then abscond to Turkey with 1.2 million Euros of his funding); he opens a shop on Phuket beach to rent out 30 scooters to tourists – which eventually morphs into a business in which he rents out 70 high-powered motorcycles.

Julien Vincent does not seem to believe he will ever be caught – and yet he is already being watched. His postings are public, after all – and so the Swiss watch companies he's ripping off on a daily basis see the increasing batch of photos and grow ever angrier. Thankfully, they have investigators on the job – and the one who matters most here is Thierry Charles Dubois.

Thierry really doesn't want to be spotlighted in the first place – he puts a tremendous value on personal privacy – but he spoke with me just the same. His manner rarely betrays either his Swiss heritage or his role as an enforcer of rights, perhaps because he inherited his position from his father, who patrolled the illicit watch industry here in prior decades – thus providing Thierry an intimate education in this unique milieu – in the clone business (he also has a law degree from Syracuse) – as well as in China itself. He finds the cat-and-mouse chase invigorating, sure – but also funny, a bit silly, never-ending. Grown-ups playing cops-and-robbers.

Thierry wears jackets without ties, smart spectacles perched on his nose. There's something familiar about this mien, like you've encountered it before. He is frequently bemused by the memes posted on Reddit by the counterfeiters themselves (who may or may not know that he's watching and grinning; a lesson: there's always somebody watching in this watch game – atrocious pun intended).

He executes his dual role with a (legitimate) 42mm olive green Patek Philippe Aquanaut chronograph in white gold on his wrist – reference 5968G-010, a six-figure timepiece he wears very discreetly and that I spotted only because I was hunting his radiocarpal joint for just such a piece (#journalism). In one role, he serves as Managing Director Asia Pacific for the Fédération Horlogère, the official Swiss watch trade group. In the second, overlapping gig, he labors to hunt down fakes and their creators on behalf of individual corporate clients – from Audemars Piguet to Cartier to Rolex – under the company moniker "Selective Trademark Union."

Here, in the second strand of his public-private work, Thierry encounters a problem – he wants to grow STU so that it plays a larger role in global counterfeit luxury trade enforcement, but he detests the idea of exposing his actual self any further. He desires anonymity – out of a certain demureness but also necessity – it helps him none if the bad guys recognize his mug.

There are details I cannot disclose here except to say – you don't want to challenge the STU – they are very dogged and very precise – all this despite Thierry's sincere appreciation for the quality counterfeiters manage to obtain in their work – he truly does believe their work to be high-level horology, true watchmaking (it just happens to be verboten).

* * *

So how does a Chinese government bust of a counterfeit watch factory ever come to pass, considering the criminal industry's embedding within a legitimate watchmaking region? What reason does China have to care about these crimes until big brands begin abandoning their market, which no luxury marque can afford to do? When do the Chinese decide to move in on an actual economic sector in which they should take a sort of perverse pride?

It's really all lobbying.

When Thierry's individual watchmaking clients decide they're fed up with a particular Chinese replica watch (the euphemism for a counterfeit), or a single rep-making operation, Thierry then has to work to enlist the Chinese authorities to set up a sting/raid/bust. It's a very delicate relationship, and Thierry must encourage the Chinese to intervene while remaining on good terms with their not-always-easygoing bureaucrats and law enforcement (there are two branches he deals with, one of which stages raids that result in minor punishment — slaps on the wrist; the other metes out rather more severe penalties; Thierry does not get to choose which one takes the lead).

Thierry is a man who collects incriminating info bit by bit and then intercedes and makes a formal request when his clients feel it appropriate. Weirdly, a major Swiss brand once had no issue with a clone of a six-figure chronograph composed of a most exotic and difficult-to-manufacture material— despite its maker, APS, having executed a startlingly-gorgeous and finely honed replica version. The fact that APS is nevertheless raided 11 days before this book's release, after I've finished writing it (or so I thought), probably means AP started caring after many years. After countless meetings. A great deal of imploring.

So Thierry must lobby his own clients to care also in a sense. He is the go-between pleading for sanity from the Swiss and Chinese both. Thierry has no legal authority on the Chinese mainland or in his office's site of Hong

Kong, though, obviously (in Switzerland, his clients at least empower him to present his findings with gravity).

So he *must* convince the Chinese officials to conduct a raid from a kind of sideline position. That is a major effort — incentivizing the big raids even once the evidence of bad behavior has been gathered and turned over to the proper authorities.

But that's all Thierry can do — collect comprehensive records of wrong-doing and then lobby the Chinese to take them to heart. Those are not easy conversations — he often feels stymied. But with patience, the work pays off. And once the Chinese authorities agree to move in, they do utilize cutting edge tech to home in on the cloners. The Chinese government's power to find a single human, or even a loose network of them, should never be overlooked.

It's like Big Brother finally falls into a major watch hobby. And he can tap into his vast troves of data to root out the groups suddenly deemed worthy of attention. It does not help counterfeiters that they use the Internet and cell phones in a country whose telecommunications are filtered through state surveillance mechanisms of the highest order. That is about as specific as sourcing allows me to be. But there is no tech or methodology that can allow the violator to evade the state's watchful eyes (once those eyes begin watching) for very long.

* * *

But also, Julien Vincent makes STU's job so much easier: Whenever he addresses his French helpers in the business, he calls them on his monitored social media "l'équipe" (the team) or "les amis." It's an obvious pattern - and when he tells his team on those platforms that he is going to Guangzhou to check on the supply chain and inventory, the STU knows it has a big break:

Julien Vincent is leading them to the highly-skilled creators of the Rolexes he sells - at a point in time when the Noob factory is still hidden from authorities, its cover un-blown, its details unknown.

Julien will change all that. In his defense, when he arrives in the Chinese province, he does not travel himself directly to the Noob factory for inspection. He isn't that dumb. He instead meets his partners, his associates, in a coffee shop.

But as Julien has posted online of his trip to inspect the goods, anyone trailing him knows very well: those with whom he meets are likely themselves to visit the factory, to be involved in the enterprise of Noob. So *they* can be tailed from the coffee shop to the production facility. He's *that* dumb.

The result is that the Chinese government, tipped off by the Swiss enforcer, is shown a trail leading to Noob. Easy as that (except that it took countless hours of surveillance and tailing by the watch investigators in the first place).

One French 20-something living large brings down an empire of watchmaking with mechanisms constructed by a heritage watch brand in business since the 1950s. Mindbogglingly self-defeating.

The Chinese security apparatus labels the April 2021 bust the "Cloud Swords" crackdown. In all, 17 sites of production go dark – suppliers, subassemblers, and the main Noob facility itself. Every Noob leader is arrested. The operation is valued at $42 million.

The sentencing comes in July 2023, slightly more than two years later: 15 Noob higher-ups who have been detained face a judge – and the heads of the crew receive more than the usual Chinese slap-on-the-wrist for counterfeiting (although others get off rather lightly).

Xu, the leader, gets six years and six months. Zhuang three years and one month (recall, he pulled out of the scheme two years prior to the bust). Others are handed down prison terms of four years, one year, and even mere suspended sentences in some cases.

Whether any of this is enough to deter future counterfeiters (six years is the max?) is a question only potential Chinese successors to this operation can answer.

But the Chinese bust is hardly the end of the Noob story – for the raids have no bearing on the freedom or prosecution of Julien Vincent. All he has done is visit a coffee shop in China – and as he is domiciled in Thailand as a French national, he is an international conundrum anyway – who has the authority to prosecute him and for what? He hasn't actually ever been caught making a fake watch or personally selling one – so what's the best legal strategy to nab him?

The answer arrives not through any legal strategy at all. October 2022, Koh Samui: A year and a half after Noob is brought to its knees, Vincent is riding a scooter with an associate when they get into it with a Thai man on a motorcycle. Words are exchanged. A brawl ensues. The police show up – and Julien Vincent, international watch trafficker, is arrested for assault.

This is such a tremendous liability for Vincent that it's amazing he ever risked any kind of arrest in Thailand for any offense at all. By October of 2022, Vincent knows well what has happened in his native France over the summer. His entire distribution network has been dismantled by that country, and his second-in-command, the aforementioned Florian Rebani, imprisoned (a full circle moment: Rebani apparently only entered the illegal watch trade after being sidelined years earlier in a motorcycle crash).

Vincent, in other words, knows before his road rage incident, however blinding that rage might be, that France is seeking his extradition. Thailand isn't looking to pick him up, necessarily – but should Vincent find himself incarcerated by the Thais anyway, they'll likely have no issue handing him over to his home country for prosecution.

And so they do – Vincent is returned to France in early 2023 – jailed but stubbornly defiant. He tells the press his wealth is increasing even as he sits incarcerated – after all, he's invested in cryptocurrencies: "I watch TV and see that bitcoin is going up," he says. "I'll be out in five years. I'll have everything I need."[142]

A few days before the start of his trial, Vincent fires his own lawyer; during the proceedings, he goes out of his way to say he doesn't "give a damn" about the court case; he also refers to a prosecution lawyer as "a clown."[143] The trial takes but two weeks, ending in April 2024 – almost exactly three years after Vincent's carelessness led to Noob being raided. Boasting a thin mustache, Vincent wears sunglasses in the docket on the day of his sentencing.

Rolex having lent its legal team to the prosecution, the man is punished heavily, at least from a financial perspective.

He receives but 4 years and 6 months in prison, but is fined record sums: to the French customs authority that his shipments skirted, he owes a whopping $220 million. To Rolex, he owes an additional $2 million penalty.[144]

And though he primarily counterfeited Rolexes, as he dabbled in the production of other brands, he owes them compensation as well:

Audemars Piguet: €50,000
Patek Philippe: €50,000
Richard Mille: €50,000
Omega: €50,000
Chanel: €50,000
Hublot: €44,000

***

An epilogue takes us backward, two years before Noob is ever raided, one year before the Pandemic traps us all at home. Takes us right inside that meeting after the Noob procurement man Zhuang voluntarily leaves the business. Now Chen is taking charge of finding a replacement chronograph mechanism for the cloned Valjoux of modest quality. And Chen, all ballsy, petitions Liaoning Peacock in the city of Dandong. Asks one of the most historic Chinese watchmakers to produce for an underground network of dealers a chrono of highly technical specification – a clone of the Rolex 4130.

A clone 4130 movement I bought from China during the writing of the book.

The story I pick up from underground sources on the Web who have rather convincing sourcing for their narration goes like this:

Peacock doesn't just accept the Noob offer because they need the work, illicit or otherwise. They accede to the request because they plan on marketing this cloned movement to any company that approaches them with a desire for OEM services – anyone who wants Peacock to build for them a chronograph silently under a third party brand name.

It's a very logical legend – think about how much business Peacock truly believes Noob will deliver it – they are counterfeit peddlers at a high level but peddlers all the same – and there is considerable risk involved. So why not use the movement in as many legit-market watches to make the risk more palatable — more financially worthwhile?

Thusly, does Peacock decide to market its Rolex clone to third-party brands as the SL-4510 movement. One insider posts to Reddit a supposed screenshot from the small period of time when the 4510 is actually for sale:

SL-4510

中心时/分/计秒针指示；
Φ30.50mm, H:6.50mm；
3点位：计分（30分/圈）；
6点位：小秒；
9点位：计时（12小时/圈）；
自动兼手上链/止秒；
频率：28800次/小时；
Center Hr/Min/Chrono Sec hands;
Φ30.50mm, H:6.50mm;
3H: Chrono 30 Mins;
6H: Small Sec;
9H: Chrono 12 Hrs;
Auto & Hand winding/Stop Sec;
Frequency:28800 Vibrations/Hr;

But Peacock flies too close to the sun; here, Chinese watch market experts can only speculate as to the fatal mistake and I can't confirm which chronicle is the most accurate one.

The most widely-spread story is that Peacock puts this movement on display at a watch fair. And Chinese watch and clock fairs are swamped with legal teams and private investigators, every Western company monitoring the goods for any kind of wrongful design, anything that infringes patent or trademark.

So the movement sits there on Peacock's table, presumably under bright exhibition lighting, until word reaches Rolex – likely through some lawyer at the firm Deacons – that a cloned 4130 is being offered to third parties. Deacons sends a very stern letter by way of its lawyers in Hong Kong to the panel at the watch fair, so the watch fair says, *Shut it down.*

This is just one version of the tale — but they all hit upon the same points: Rolex being made aware of Peacock's doing, and Peacock ceasing all presentations of the SL-4510.[145]

And after the Noob raid not long after, it modifies the 4130 movement slightly (it does not have to be altered much, as the 4130 is now no longer protected by the 20-year period of its patent, which was filed in 1990). Peacock basically makes one change – restricting the bi-directional Rolex rotor and making it uni-directional.

Thusly, does Peacock absolve itself of IP theft, wash its hands of a sordid affair involving crazy fines, Phuket beaches and a procurement pioneer named Chen. They confine a rotor to storing up potential energy only when it spins clockwise (if you are looking from the back).

They call this *new* engine the SL-4801. Neither Rolex nor Peacock returns my messages about the affair.

The fakers have been sentenced, the factories raided, the movements renamed and rebuilt just within the boundaries of the law.

There is the sense that there is nothing more for anyone to say anyway. The truth is already spoken in the maneuverings, the negotiations: The line between clone and legitimate architecture is a precious thing. Only very powerful bureaucrats get to decide on what side a Guangdong province watchmaking crew — or a caliber manufacture in Dandong working as a supplier — falls.

Further proof: A legit Guangdong watchmaking microbrand — a group of youthful manufacturers who are the direct ideological offspring of the first Shenzhen entrepreneurs — named San Martin now sells under a luxury subsidiary brand a chronograph...with the Peacock movement only slightly modified from the Rolex one that everyone has agreed is fair game. The SL-4801. The watch truly looks like a panda. I bought one — this hybrid in the metal of Swiss ideas and Chinese mimicry on the very edge of the permissible. For a piece whose great import derives from the collision of its Eastern and Western IP, it's disarmingly, smilingly adorable:

Of course, I ask Peacock itself in 2026 if it wanted to refute any of this reportage: "How did the Noob-related raids of five years ago affect the morale at the factory?" I write to the company. Which raid Peacock could state did not include its own facilities. And I add:

> *"Today, the most eminent and accomplished clone calibers emanate from Dandong. To what degree are you aware of their provenance, and are those very well-engineered movements achieved surreptitiously possibly with equipment Peacock has procured?"*

To which query, Peacock could reply, It's not our equipment making these clones (let alone us — I gave them ample wiggle room). But they deny *nothing*. What they answer instead:

> *"Regarding the specific historical market enforcement actions you mentioned and their impact on the industry, as well as rumors surrounding the sources of unauthorized movements circulating in the market,* ***Peacock Watch officially declines to comment.****"*

Even the finest superclones are sold and shipped to customers enveloped in a confoundingly clingy shrink-wrap, and sealed further inside a plastic zip-closed baggie (as if narcotics). This package bears the counterfeit factory's logo and/or the genuine model's reference number and/or the dealer-issued order number (plus, in the case of the VSF pouches above, the version of the model sent; on the left, for instance, "V3" indicates the third iteration of a particular Rolex chronograph.

10

The Body Keeps the Score

January 2002. A delivery truck picking up polished gold Rolex watch cases, just outside the gold polishing firm Miranda, is held up at gunpoint by two armed men and stripped of all its glistering wares.

The robbers drive off with the van and a hostage, but they eventually ditch both and keep only the gold watch cases.

Or maybe what the van contains and what is taken are only the polished gold backs of watch cases and not the full cases themselves.[146]

The eventual (slow) criminal investigation leads to perhaps the greatest Swiss watch industry scandal of all time – or ample evidence that the industry cares more about protecting itself from damn good clones than it does about integrity as a concept. For reasons only he knows, Swatch cartel CEO Nicolas Hayek chooses to go with the first narrative: "It is as if the cardinals in Rome had emptied the Vatican's vault," he says.[147]

* * *

No one who really knows Jean-Pierre Jaquet is ever taken by total surprise when his name is connected to unsavory happenings.

Jaquet is an icy man, hard, sullen. His father ran a pawnshop (or a kind of second-hand goods operation akin to one but slightly shabbier). As a boy, Jean-Pierre taught himself the art of machining very fine watch parts –

he was schooled only during the winter months – and that contributed to his belief that he was in some constant battle against the very industry that provided him work.

He started his professional career restoring old watches in 1976.[148] His modifications of mechanisms were performed in almost domineering fashion, as if to say, *Look at all you so-called haute horology houses – look at how you need this poor, uncultivated and informally educated mechanic so.* This demeanor, not to mention his unchanging mien, a stern, unvarying, unnerving visage, in turn, led industry peers to call him "The Pharaoh."

In 1980, he was handed a 24-month prison sentence for receiving/handling stolen goods.[149] He was also rumored to have passed off counterfeit watches as the real McCoys during his '80s period as an antiques dealer/-pawnbroker himself and was reputed to have participated in the major Italian market for top grade fakes.

In 1992, Jean-Pierre Jaquet incorporated a company he'd purchased two years earlier as Jaquet SA. Based in La Chaux-de-Fonds, Switzerland, this corporation took base ETA/cartel movements, modified them so they were capable of great complications and sold these added-value high-horology engines to the top brands, including Franck Muller and Bvlgari. Within a decade, the company employed 180 people and had annual sales of 44 million Swiss francs.

Perhaps Jaquet went straight. But Jaquet the man cows the engineers at his own company by imperiously and brusquely solving small mechanical issues their bright minds and able hands simply cannot.

The Pharaoh engenders many fearful whispers with his utter disregard for Swiss watch industry protocol.

Jaquet finds himself in no immediate trouble following the January 15, 2002, robbery of gold Rolex cases. Here is a man well-known for at least some participation in shady dealings – an ex-con and watch dealer. But he proceeds to go about his business as per usual.

Earlier, in June of 2002, a machining firm in Le Locle used by the watch industry, RSM, was robbed of at least 10 kilos of gold at gunpoint.

But that case also goes nowhere until the following year when an insider at Ulysse Nardin steals gold timepieces worth half a million Swiss francs in total. This domino knocks down several others – the last falling at the feet of Jaquet, whose company is finally raided in October of 2003, or 22 months after the Rolex watch case heist.

It almost appears as if nobody investigating the Nardin case ever meant to encounter Jaquet. It seems everyone is led to Jaquet by dint of circumstances that keep getting stranger and further beyond their control. And it seems the major watch companies do not want to cooperate as fully with authorities as they publicly state.

Finally, in October 2003, 24 police officers with submachine guns enter the Jaquet office and seal it off. They use the guns to keep employees from nearing the facility as they search it; they even block off the access road from Biel.

Jean-Pierre Jaquet is arrested at his home.

He is thrown into solitary confinement, accused of masterminding a counterfeiting ring that places parts and movements stolen by Jaquet (and like-minded henchmen) inside authentic and stolen gold Rolex cases.

There is no verdict for a full five years – at which point Jaquet is found guilty of receiving stolen goods, watch counterfeiting, use of forged documents, and tax/financial offenses. His sentence? Just 4.5 years.[150]

That detention obscures the profoundly disturbing breadth of what is actually uncovered: that the eponymous head of Miranda, Jose Miranda – a gold polisher for Rolex – was complicit in the heist of the polished gold eventually bound for Rolex.

That The Pharaoh pressed an Italian who owed him 200,000 francs to organize the staged robbery, which wound up involving a Bosnian hitman.

That Jaquet pressured workers at watch suppliers and at Ulysse Nardin to steal components for his operation. In fact, the counterfeits he then created were so good they became known in the Francophone press as "true fakes."

That Jaquet had possession of Franck Muller brand watch materials that indicated he was actively counterfeiting that brand's goods (he claims he was simply holding onto various prototypes from the work he had done legitimately for the brand over the years).

But this is where it gets so strange. In Switzerland, a criminal prosecution affords victims the opportunity to join the action as civil litigants instead of pursuing a different case entirely. The question of dropping out is no small one. Franck Muller does *not* pursue civil action against Jaquet as part of the case even as the state insists Jaquet is counterfeiting Muller pieces and implicates Muller's uncle Gaby as working on behalf of the larger Jaquet gang as the insider – the supplier of legit parts.

Ulysse Nardin also ceases pursuing Jaquet in court. Its own CEO says the temporary branding of the case as the "L'Affaire Ulysse" is more damaging than the actual heists.

In the end, certain brands care less about restitution or justice than the revelation of Jaquet's achievements. An unschooled machinist, so long as he has the raw metal in his possession, is able to create clones just as brilliant and exquisite as their legitimate counterparts. So long as he has the right hands, the correct dials – he can bring to life any brand's complicated watches far outside its purview, reach, factory floor.

Jaquet has essentially demystified once and for all the Swiss watchmaking process – demonstrating that it is simply a matter of supply chains and networking and skill. Yes, he has to infiltrate those chains to run his operation and the careful compositions reflect an autodidact's unique eye and expertise.

But what is a Swiss watch after this trial other than the careful choosing and integration of componentry-makers and a thorough session of quality-control thereafter?

If Jaquet can do it, can't anyone with the means, will, social engineering?

The court does find that Jaquet personally handled the RSM machining gold – which is worth 650,000 francs. And the Rolex cases, too.

And Rolex and Richemont (owner of Cartier, among others) do not drop out as civil complainants within the larger criminal case. The court orders a total restitution to these victims of 750,000 francs.

Except that two months after this judgment, upon Jaquet's appeal, the court frees him. He does not resume serving his sentence until a Federal Supreme Court ruling upholds it in March 2011 – or more than nine years, nearly a full decade, after he presses an Italian and a Bosnian to hold up a gold watch polisher whose very proprietor is in on the robbery.

And even then, when Jaquet resumes his sentence, his lawyer negotiates with authorities so Jaquet can work as much as possible – in the watch sector, naturally – while serving his time.

Serving his time: I think of this phrase when I consider the La Joux-Perret movement inside the Farer watch with the white enamel dial I have been wearing.

For between 2003 and 2004, following the arrest of Jaquet but long before his trials conclude, the company's other stakeholders quietly buy him out of the enterprise for 11 million francs. They rename the firm "La Joux-Perret" after the plateau upon which its factory sits.

And in 2012, just a year after the Supreme Court upholds an arguably paltry sentence for Jaquet, Citizen buys the self-taught Pharaoh's former company, La Joux-Perret (along with its interrelated corporate entities Prototec and Arnold & Son), for $70 million, paving the way for its movements to spread not to underground timepieces but to microbrand and luxury watches such as a Farer, Panerai, Hublot, Tag Heuer.

You know nothing of the Pharaoh – he casts no shadow over your idea of the piece. That was the goal of the industry from the start.

Serving his time: But I do think about him, when I consider the La Joux-Perret calibre in a simple Farer three-hand watch. A British microbrand's thousand-dollar timepiece recalling a Swiss criminal legacy — the criminal's company now part of Citizen, the company cofounded by aforementioned rebel Swiss watchmaker Rodolphe Schmid.

History written in the metal.

* * *

Perhaps *the* key to understanding the drift of horological IP from England to Breguet's France to Switzerland and then America and then Switzerland again and finally to Asia is to examine the automatic movements presently powering the devices themselves.

These engines tell the history of the industry – at Swatch Group, which is the modern corporate version of the cartel that used monopoly power to push the Americans out of watchmaking in the 20th century, the big revolution has been a reduction in vibrations per hour.

By taking its subsidiary ETA's 4 Hz (or 28,800 beats per hour) movements and lowering them either to 3.5 Hertz (25,200 bph) or even just 3 (21,600 bph), Swatch engineers have managed to eke out power reserves of 72 hours (Mido Calibre 72, Longines L.888) and 80 hours (most famously, the Tissot Powermatic 80, but there are other corporate iterations).

And yet, not only does the slowing down of an older movement not constitute the creation of a truly new device, the older movements are far hoarier than they might upon initial inspection seem.

The 3.5 Hz calibres iterate on the slim 2892-A2 – a version of the 2892 brought to market in the mid-1990s; the original 2892 was invented in 1975 but based on the Eterna 1466 movement conceived by famous cartel designer Heinrich Stamm *in 1962.* That means the architecture is 64 years old at its essence.

The 3 Hz calibre is the latest iteration of the thicker 2824 cartel workhorse engine – launched in 1971 but patterned after the Eterna movement of 1955. In other words, the configuration was first sketched out more than *70 years ago.*

In reaching back to the middle of the century, the movements date to the exact moments when the Swiss cartel was demanding control over American timekeeping companies' production by threatening to withhold repair parts unless its dictates were followed.

But that's only a sliver of the legacy of the ETA ebauches – because their very age means they cannot be protected by patent anymore without modification. So many of the legitimate economy Chinese microbrands that have proliferated in the last 25 years use the same ETA architecture – only they buy the copies legally pumped out by other companies: the Sellita SW200 and the Soprod P024 and the STP1-11. These are all totally lawful copies of the cartel's 2824 movement. And it's not just microbrands: These particular calibres abound in the sub-$3,000 luxury watch space. And the

Chinese, of course, have contributed their own 2824 copies: the Seagull ST2130, Hangzhou 6300, HK Precision Technology PT5000. Even the Rolex brand's retired T600 movement for Tudor was not much beyond a 2824 with customization.

All of which is to say, the technical workings of the watch market to this very day tell a story of a monopoly Eisenhower considered busting. And those fine watches in the sub-$3,000-range lacking a cartel-derived calibre?

They almost exclusively use a 9000-series Miyota – a 4 Hz movement created by the Citizen company in 2009 as a response to Swatch limiting its sale of 2824 movements to outsiders. Which means the only real alternative to the cartel's legacy scaffolding in the first few rungs of the watch industry is produced by a company cofounded by a disgruntled Swiss watchmaker who had a falling-out with the cartel, which refused to grant him permission to create "Swiss" watches in Japan.

That's right – Miyota is part of Citizen.

Which leaves Rolex. If that's simply a byword for luxury as opposed to a company reference — if it's a useful metonym for all posh brands – we can say those watches it represents generally do feature unique, in-house movements that go uncopied in the industry less for their complexity than for lack of demand (if not for both).

You won't find an illicit clone of the Vacheron Constantin Calibre 3655 – a movement with 41 complications (any feature beyond simple, three-hand time-telling) and 1,521 constituent parts. For one, the mechanism is prohibitively complicated and expensive to produce. But also: There are too few clout-chasing replica (fake) buyers looking to flex the Vacheron pieces in which the movement features.

There's not enough demand to merit serious reverse-engineering.

And yet, if we are talking in terms of Rolex watches themselves, whose movements afford massive power reserves and incredible accuracy even while being produced at tremendous scale (the brand retails one million pieces per year), the picture changes dramatically. Here the underground market desperately desires 1:1 copies of the catalog's offerings, and so a supply is amply and accurately made available.

Actually, no movements are studied as heavily, or produced as faithfully to the genuine articles, by Chinese counterfeiters as Rolex's – and there are many so-called factories that compete to produce the best counterfeit version of the 3230, 4130, and 3285 (among other calibres).

The ultimate facsimiles of these engines can work with swapped-in genuine parts and reliably keep time for years. The Dandong-produced 3235 in my rep titanium Rolex Yacht-Master and my rep Sea-Dweller beats robustly and on point across its three-day power reserve without fail.

Its steady, dependable rate has been, in many doctor's waiting rooms over the last year, my security blanket. It's reassuring to have a small, non-electric machine perform on point and exactly as promised amidst natural phenomena neither man-made nor predictable.

See, I've been writing this book while attending ceaseless doctor's appointments with my pregnant wife. I have been awaiting the birth of my first child, the gestation of whom has been beset by myriad complications and one shocking, terrifying, week-28 hospitalization.

I have been very scared, very often, and, trying not to show it, I have eyed the time. I have tried to settle into the comforting pocket of a consistent beat-rate. A second hand sweeping around a dial. The way a well-adjusted watch, even one crafted against the law, glides into the future even as its balance wheel oscillates side to side as if batted about by rocky and unpredictable exogenous forces.

When her water abruptly broke on a Saturday morning — 35 weeks and four days into her term and a full 10 days before a scheduled induction — I was not attired in a single item I'd ever planned on wearing in the hospital. I went, enervated and elated, as I was.

And I happened to be wearing on my wrist a VSF-produced Rolex reference 126619LB — the so-called Cookie Monster dive watch whose shimmery blue ceramic bezel encircles a profoundly glossy black dial. It's stainless steel even though the genuine article comes in white gold, a discrepancy visible only to those whose trained eyes can detect slight differences in the warmth of alloy hues.

For me, that metallic disparity has always been invisible — and is now utterly insignificant; the Cookie Monster will always be the piece by which I tracked our daughter's emergence into a world frightening and filled with possibility.

No timekeeper will ever feel more genuine.

11

All of Us

It seems folly to count the hours – blissful ignorance to abstain from their appraisal.

We are all going to die. Why, then, would we want to fritter away even the slimmest fractions of our lifespans glancing at the current time, which – always being further advanced than when we last checked, necessarily – can only really recall for us our finitude – mortality?

(Sure, we need to be on-time for events *and* meetings, but…And yes, Socrates deemed the unexamined life not worth living, but he also drank his hemlock with a placid acceptance of terrible fate – which, aroused in others, would soon doom all humanity to an even earlier demise, procreation requiring, if unequally, parts delusion and defiance.)

A watch, its hands always sweeping forward, is nevertheless counting backward to the wearer's inevitable demise. To our ends unwound – moments of final weakness, when we can no longer hold even the slightest bit of energy in coiled reserve. When our movement gives out.

Measured time mocks us with an insidious, slender-marker tick – small checkpoints along a path that seems circular on a dial but tragically linear in real life (perhaps science will solve the latter issue one day).

We are conveyed by chronology forward, by repeated daily numbers assured that we're not nearing the end – until we're nudged body and soul into the abyss – where the idea of measurement not only becomes meaningless – it's revealed never to have had any at all.

Fathomlessness.

A watch is retrograde motion, a system at war with itself – both pulling us ahead even as it tugs us back, once more into numerical iteration.

And yet we look. We can't help but check the time.

* * *

It's not that watchmaking is corrupt – it's that we've gotten here, to this point in IP history, through a series of decisions that no one seems even slightly curious to reconsider – an alternate history ignored. But what might have been had we gone another way?

What if America were the locus of the watchmaking world – what if Waltham had never been successfully penetrated by spies and an effective double-agent? What if Francois Perregaux had succeeded in selling the Swiss watches he represented to the Japanese in the 1860s – would their wholesale embrace of his timepieces have strangled both Citizen and Seiko before either company was ever born?

What happens to the greatest Breguet of all time – and all the business Breguet does – if the person for whom the piece is made, Marie Antoinette, isn't put to the guillotine? There's a watch history that looks very different, perhaps, if the French Revolution itself does.

There has never been an inevitability about any of this – neither the success nor failure of any firm. As it happens, Japan didn't adopt Western time until 1873 – making the Perregaux venture ill-timed, at best.

But something else happened that same year – a reminder that all these chains are interwoven – that the historic tapestry of watchmaking, of building up and breaking down barriers (not necessarily in that order), is a fragile, fungible thing.

Picture a customs house, 1873, somewhere along the Eastern seaboard. A crate pried open, straw packing brushed aside, and there they are – a profusion of movements bearing the fine names of the greatest American watchmaking firms of the era. Customs officials up and down the nation's ports are finding the same thing: foreign-made mechanisms engraved with American prestige, shipped in by men seeking to pass off their work as legitimately domestic.

"It is understood the government will press the cases against the parties who have been importing inferior watch movements into this country with forged names of American manufacturers engraved on them," writes The Washington Chronicle, a DC paper, regarding the final decision taken.[151]

But the news is viewed as trifling, a matter insignificant – the mad men on the Rolex account have yet to convince the world that a watch is anything more than that (but boy, will they instill in the global psyche the strange notion that a Rolex is unsurpassable...soon).

So yes, the Washington Chronicle reports the news – America will go after the counterfeiter of its best watch movements. It will protect itself.

But the item runs below a much bigger story about the way overcapacity in the tea market has led several insolvent tea firms to reorganize – has led over the prior two years to a loss of $6 million in the industry.

That's at least $160 million today if you account for inflation.

One hundred years after the Boston Tea Party, that one comestible, caffeinated commodity is still more crucial to the nation's psyche (and economy) than anything to do with luxury watches.

The scariest thing about the history of watches is what it says about the flittering nature of value itself.

* * *

Nature abhors a vacuum, yes? The hole gashed into the market by the shuttering of Noob is soon filled by the emergence of Clean, a counterfeiting operation that incorporates most of Noob's horological innovation and artistry. It, too, is then raided in the early months of 2025 – an operation that nearly rivals the Noob-swatting for the largest counterfeiting bust in Chinese watch history. Twelve factory sites are shut down — plus an electroplating plant with thousands of crowns.

V7 Factory, which makes particularly good clones of the Gerald Gentadesigned IWC Ingenieur, is raided not long afterward.

By the time Clean is taken down, I already have purchased through trusted dealers a Rolex Explorer and Oyster Perpetual from the operation. They are both fine watches, with the former keeping tremendous time, despite arriving from China with a movement filled with grit and grime and hair.

This is not uncommon. The trusted dealers offer photographs and videos of your exact watch during a process called "Quality Control" – including photos of the replica watch's numbers on a timegrapher (its beat rate, amplitude, daily deviance – is it five seconds slow or fast daily?) – but they don't provide the sort of microscopic view one would need to spot the fine dust that can fall into minuscule movements during assembly.

And the reason is not sloppiness. In an aboveboard watchmaker's facility, assembly takes place in an air-filtered, pressurized cleanroom one can only enter wearing protective equipment much like what proliferated during the pandemic.

Your body is covered in a "bunny suit" – one made of continuous polyester filaments interwoven with carbon fiber (the idea being that the carbon fiber's conductivity will repel charged dust).

The T1 room (the "T" stands for "terminage") is where the movement is assembled. T2 is where dials and hands are added and then the whole unit is encased. These look like hazmat spaces, if you watch people coming and going – as if what's being dealt with in the rooms is either a viral outbreak straight from a Michael Crichton novel or the first alien life form to touch down on this planet.

And the reason for all these intense precautions is that, snake-like in some small sense, we are always shedding skin – subject to microscopic flaking we can't see with the naked eye but occurs just the same – meaning we're never quite the same humans from one moment to the next.

Counterfeiters are loath to shell out for such filtered and pressurized facilities when they're liable to being raided and shut down or forced to scurry to a new warehouse with little notice. They don't invest in them, at least not in the air controls (there are even air showers legit workers take, which blast them like reverse suction Dysons).

The result is that no single space in the counterfeit world is more obviously bound by rigid rules on covering up than any other. It's all one expanse. And so that part of us we are always leaving behind – shed skin particles invisible to the naked eye but obvious through a loupe – can fall down – like snow upon all the living and all the dead – into the heart of the clone watch's movement.

It's both a poetic reminder that the replica watch is hand-crafted and also a major problem for the mechanism's longevity. Basically, the watch can keep perfect time until it suddenly breaks if you let the grime sit; if you service the movement right away, and it was a robust clone to begin with, it can last a goodly number of years – but a full servicing generally costs the same as (or a major fraction of) the watch in the first place, and many ordinary watchmakers are wary of working on clones for fear of being blackballed by brands when they next order replacement parts.

* * *

The Chinese cannot control the free-market processes they set off in Shenzhen, which spread to Guangzhou – all over the entire watchmaking province of Guangdong. In the 2020s, Xi Jinping introduces his concept of Dual Circulation – two policies for how companies should work. One focused on building up the domestic market to be culturally Chinese and self-reliant. The other focused on pleasing the very different consumers of the international market. Never the twain shall meet. These policies are impressed upon local watchmakers by the slightly hoary, fading Guangdong Province Horologe Association (founded in 1992) and the Shenzhen Watch and Clock Association (founded in 1988).

From the top-down, the government makes clear what it wants: legacy manufacturers who specialize in OEM – in making quality but economical products for other brands, wholly anonymously – should begin creating their own consumer-facing brands that indulge in Chinese mythos, legends, stories. Brands that represent the culture of the people all around them.

Zhu Shunhua, president of the Shenzhen Watch and Clock Association, stresses from his rostrum and in official correspondence constantly the idea the Chinese brands all around him must exploit "Guochao" (the national trend, is perhaps the best explanation of this pervasive communist term) to tell Chinese stories that build up consumer confidence in their own heritage.

There's actually a kind of unexpected self-doubt Zhu's remarks and others inadvertently betray: The Chinese are well-convinced in their technological prowess – they don't really have any doubts at all that their products at nearly every level of the market can compete with those of the Swiss or the West.

But the moneyed Chinese are keenly aware of just how artificial and top-down, imposed government culture is – they don't want their own companies' goods even when they outperform foreign competition because the storytelling isn't there – there is not a shred of the heritage even a Japanese Seiko arrives embodying:

Did you recall that after the great Kanto Earthquake of 1923, when the brand then going by Seikosha and located as ever in the Ginza, Tokyo, suffered tremendous fire damage – such that the pocket watches in the shop undergoing repair and regulation for customers became attached one-to-another by the flames, as if in an art exhibit? Glommed together in a charred, twisted mass until they had achieved a ruin-porn effect – an aesthetic born of incendiary tragedy:

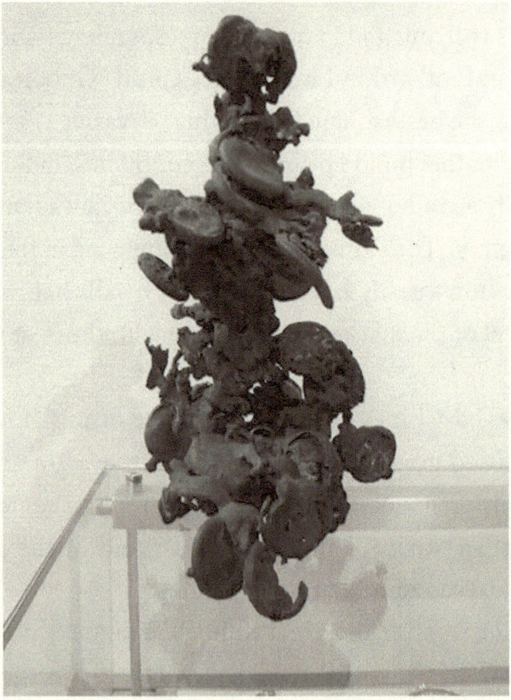

And did you recall that Kintaro Hattori, proprietor of the brand, then took out a newspaper ad to promise all 1,500 customers whose timepieces had been under repair that he would personally and free-of-charge give them all entirely new watches?

Not only did that ad run 102 years ago – I actually asked the current president of Seiko, Akio Naito, if I could look at it during one of our many online instant-message exchanges over the course of the spring of 2025. And he just sent me the file:

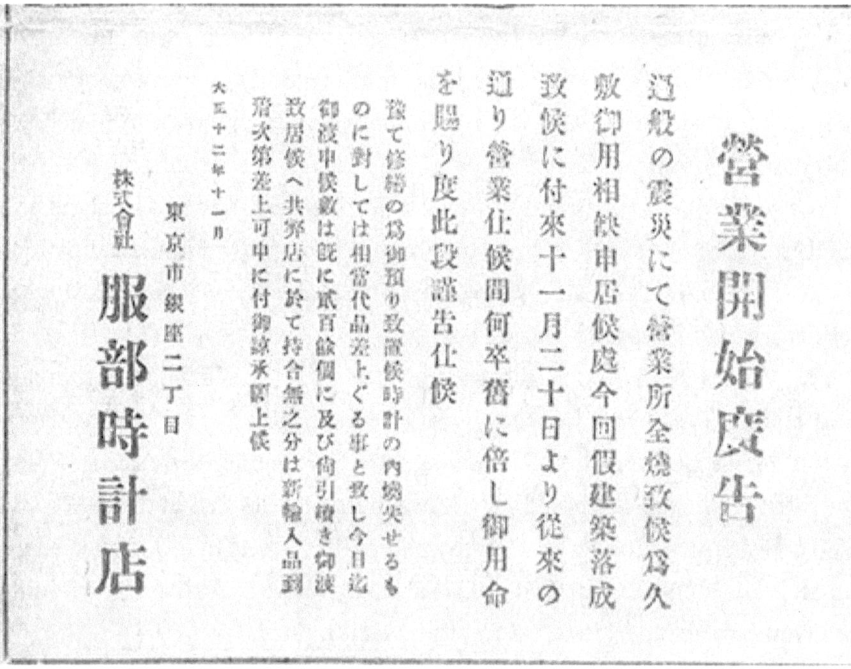

營業開始廣告

這般の震災にて營業所全燒致候爲久敷御用相蒙申居候處今回假建築落成致候に付來十一月二十日より從來の通り營業仕候間何卒舊に倍し御用命を賜り度此段謹告仕候

猶て修繕の爲御預り致置候時計の內燒失せるものに對しては相當代品差上ぐる事と致し今日迄御渡申候數は既に貳百餘個に及び尚引續き御渡致居候へ共寡店に於て持合無之分は新輸入品到着次第差上可申に付御諒承願上候

大正十二年十一月

東京市銀座二丁目

株式會社 服部時計店

The Chinese are well-informed — they know not one of their watch brands has a story as emotionally compelling — at least, so far as they're aware. The government would like local brands to lean into whatever stories they can muster evoking Chinese history and ideas.

At the same time, there has been the strange imperative for the Guangdong watch brands to do business outside the country – but only with the countries clearly on China's side in its trade/culture/superpower showdown with the US. Watchmakers in Guangdong have been encouraged to show their wares at fairs held in Vietnam, Saudi Arabia and Russia.

That is the official method of bifurcated business the government prefers. But on the ground, reality looks nothing like the mandate. The internet has brought Chinese watchmakers in the microbrand space – San Martin, Watchdives, IXDAO, Cronos, Addiesdive – closer to Western watch enthusiasts by letting them do the simplest thing in the world: talk. There are entire Reddit threads, long and well-informed, in which watch creators from those brands discuss with their American fans what dials, bracelets, casings they'd like to see in future models.

The best of these watches run well-regulated Swiss movements or the Japanese Caliber 9000 series created far more recently by movement maker Miyota (a subsidiary of Citizen Watch named after the town in which the company is based).

As a result, the microbrand watches of the Chinese – and of England and the US, which are experiencing their own horological reindustrializations but on a much smaller scale – keep better time than most mechanical watches in history ever released – period.

The Miyota 9015, first released as recently as 2009, is brilliantly thin and only about 15 years old. I've seen it not lose a second a day for weeks (maybe it was a trick of the mind). The Swiss movements microbrands use – say, the Sellita SW200 – are based on ETA calibers first devised in the 1960s and 1970s and now no longer covered by patent (the Sellita in question is based on an ETA 2824-2 from 1971-1972) and keep time as well as the Miyotas, in my personal experience, even if they're clunkier.

Of course, the Chinese brands connecting with US customers – I literally reach out to the owner of Watchdives to discuss this book and end up purchasing a pair of prototype watches for a model not yet released directly from him – can't do anything without inherent tension with their government and ours – their shipping has to potentially go to a third country before it hits the US to avoid massive tariffs – or they must defray the cost in another manner.

(The Watchdives proprietor Jassy Hu I much admire for his bluntness: he is looking to build a $2,500 watch qualitatively for $500 with his premium line – he knows all the best suppliers, and he expects the closer he gets to his goal, the smaller his margin will be and he's fine with that.)

The strange triangulation is not unique to shipping – there is a cultural bridging required, too – look at San Martin, which in part to fulfill the order to go more upmarket and more locally-devoted, created a sub-brand for the Chinese market called Jianghun, which translates to "Soul of a Craftsman," that has employed on several models the highest quality machine guilloché technique (essentially, a carved and interlaced pattern whose light play dazzles in concentric circles or other similarly eye-catching arrangements;

guilloché is one of the ways brands are encouraged to appeal to domestic buyers in a fashionable manner).

And yet this movement toward a core competency far beyond what the Swiss once dreamed possible, however Chinese culture-inspired (a chiming watch at an entry level price?), can prove an irritant in locales such as Schaffhausen, a Swiss town on the German border known for its wine, the wending Rhine River that passes through it and its major hometown horological manufacturer – IWC (International Watch Company).

My deceased grandfather, survivor of a Nazi death march who rescued another man who'd fallen to the ground and could march no longer (the destination to which he carried the man, their iniquitous, enforced finish line, was a camp, of course – Mauthausen), only possessed in his post-war years in NYC, when he oversaw the floor of a zipper factory in the Garment District, a single item of real material luxury – his IWC.

And despite how closely the watch's geographic origin on the border of Germany sounded like the camp in which he was confined and debased, my grandfather always referred to that IWC as a "Schaffhausen" – to the point that I didn't even know IWC was the corporate moniker until I was like 20.

What I didn't know at 20 – and what my grandfather certainly didn't in his time earning the sweat equity convertible into that timepiece – was that decades later, I would watch IWC/"Schaffhausen" come after a Chinese microbrand I was already writing about for my own line of work – for *this* book.

Sometime in 2025, perhaps as early as late 2024, IWC approaches the Chinese brand IXDAO (pronounced "Nine Dao," meaning "nine paths"), its production center in Guangdong (home province of Shenzhen), with a threat:

Stop making your integrated-bracelet Ingenieur watch, which takes the name of one of our watches and looks and feels in every way exactly like it.

Mind you, this is no clone – IXDAO uses its own brand name, it never pretends to purvey IWC timepieces. But it is so close to the real deal that experts classify it as an homage-clone — a "clomage."

How close legally can a clomage get?

I'm not an IP expert, but for the most part, in the watch world, such watches are generally considered sincere flattery and rarely viewed as commercial threats. These are watches priced with three digits — not five. And there's no law that truly protects the design language of a watch – especially not one like the Ingenieur, whose original Gerald Genta-scribbled outline IWC has tinkered with incessantly over the years.

The only real legal issue would have seemed to be the name – if IXDAO had stopped using the French word for "Engineer" it could have kept producing the mirror-image Swiss watch (for the movement is not copied, and that's the lone other truly patent-able component, if we ignore small oddities like clasps).

And yet, it seems that IWC is pissed about the quality of the copy, legal standing or not. In fact, by being so angry, it seems to admit that the far cheaper clomage advances almost asymptotically close to the original.

Here's what the co-CEO of IXDAO Zhang Yuan tells me about the affair, when I ask him why IWC chases his firm down instead of another, in a microbrand world where clomage watches abound (most without fuss). Mind you, I get the sense from Yuan that he is a relaxed fellow with a calm perspective on such tiffs (he formerly ran a car dealership and a pawn shop in which he offered short-term loans backed by watches as collateral – he has dealt with the gamut of humanity, it seems; he won't be spooked, it seems).

So why his clomage – was it simply too good for IWC to abide its existence (and few-hundred-dollar price for visually identical quality)? "To be completely honest, we do not have a clear answer," he emailed me, continuing:

> *We are aware that other microbrands have explored similar design directions, and we have not publicly seen the same level of action taken against them. However, legal strategy is complex, and every brand's situation is different...What we can say is that we respect the fact that established brands have a responsibility to protect their intellectual property. IWC's legal team is simply doing what they are mandated to do...As for why us specifically — perhaps our execution drew more*

attention. Perhaps our quality was too good. That last part is, of course, said with a smile.

* * *

In 1873, the Japanese adopt Western time. But the rise of individual watchmakers (as part and parcel of a microbrand world in which single-minded entrepreneurs can and do rise above the system, however rarely) means that in 2011, the then-unknown mechanical genius arrives at the premier watch fair of its era – Baselworld – with something gasp-inducing:

A wearable *wadokei* – literally, a Japanese-style watch that keeps the old Edo time of having six daytime and six nighttime hours of varying lengths according to the season.

Consider that the Edo system of varying hour lengths was abandoned before the pocket watch ever had a chance to migrate to the wrist. What Kikuno has done is iterate a superannuated timepiece in the tiniest of spaces — make a mechanism whose every component was designed to be larger and to do so with rare materials.

A horological star is born – and soon I just have to know everything about Kikuno – and at this stage in my own life, I am not even writing a watch book.

But I watch his YouTube videos, the ones in which you can see him work. I force girlfriends to watch them who share none of my enthusiasm. A single man has revived an entire system of keeping time – and a smart one, one that acknowledges just how very different time feels season to season in those countries exposed to early light and early darkness, to sowing and reaping both.

And then I get to interview him for this book – and his answers are everything you might hope for from a man who still works by himself crafting watches that do things that defy the linear progression watch history might at first seem to be.

As a kid, he loved Lego, he studied the manuals of cars to understand their assembly, his father worked at a firm that processed tungsten, but rather than further kindle his interest in creation, the job proved a bummer: he hated that his father had to work alongside others to craft an object. Masahiro was enthused by making things alone, by himself, caught up in his own thoughts alone.

Worse, the age of Japanese electronics, when the Sony Walkman ruled the world, when his native land's technofuturism seemed our global collective future, further depressed Masahiro – he felt his kinship for an older way of production, for a single person making things in a traditional manner, was already becoming obsolete.

He fell into a kind of melancholic video-game-playing routine. "I stopped creating things and escaped reality," he tells me.

But of course, his prognosis was wrong, our collective ideal of mechanized production never killed off handicraft; the digital only made the single-person, manmade analog thing that much rarer and more desirable.

"It is precisely because of mass-produced watches that independent watchmakers' creations shine and a market where they can thrive exists," he tells me.

But the best thing about Masahiro Kikuno is that he's a cinephile who quotes "Full Metal Jacket" and "Predator" and has a sense of humor about nearly everything. When asked about legacy, he says, "I am not attached to the idea that my work will remain forever…Everything will eventually disappear sooner or later."

Then he tells me a story of a Japanese man who, to avoid inheritance tax for his family, was temporarily buried with a Van Gogh and Renoir (or who requested such a burial).

"Please put my work in your coffin," he says. "However, be careful of grave robbers."

And the message he wants his art, which is horological in nature, to convey:

"There is a stereotype that watches are for telling time, so I think few people understand my intentions. I was disappointed when I was young, but I feel that people have come to understand…

"Things that are easily understood tend to be quickly forgotten."

I ask him – what if someone arrived on the scene in the future with all of Masahiro's genius but none of the manual dexterity – he could conceive of amazing devices but he could not hand-craft them. Is that person at risk of being rendered irrelevant by AI-powered robotic production?

"If you can become important *to someone else*," he says, "you can survive."

The watch — perhaps an object through which we can connect one to another, no matter that it fits only a single person's wrist at any one time.

* * *

Recall that the Swiss trade envoy responsible for keeping counterfeiting under control once upon a time – Raymond Probst – a man who wound up encountering a good deal of political resistance in Hong Kong in 1973 — winds up six years later being the main conduit for American-Iranian negotiations during the hostage crisis – in part because, as always, the Swiss are ostensibly neutral.

In fact, Probst benefits from the short historical memory politicos tend to possess – because if the Shah's overthrowers or the United States citizens they hold captive or any diplomats even the least bit involved have any sense of a much earlier Swiss man with horological industry ties on Persian ground, they might think twice about his supposed absence of bias of any kind.

The story Probst's convenient neutrality obscures is 340 years older – and it begins not with diplomacy but with a necklace.

Isfahan, early 1630s. Johan Rudolf Stadler, a watchmaker from Zurich who has already served at various European courts in territories we now call Turkey, arrives at the palace of the Safavid Shah Safi I alongside a French jeweler. Stadler rather quickly constructs for the ruler a watch the Persian wears around his neck on a gold chain.

What follows is a kind of domestic ritual – each morning, Stadler meets the Shah to wind the necklace watch, and they share wine and stories. A tipple and a timepiece. The Swiss watchmaker also repairs every broken clock already in the Shah's possession – the accumulated deliveries of courtiers and merchants. The story has been heavily researched by the author Willem Floor, and it sounds, for a while, almost cozy.

Until one night Stadler comes home from a diplomatic fete at an embassy to find an intruder – some prowler – on his grounds, in his premises. The man is a local and Muslim – that we seem to know. Whether he was looking to burgle the watchmaker's residence or he was a man in love with the watchmaker's wife and had been caught with her before we do not know (there are competing accounts). But Johan Rudolf Stadler shoots the man dead all the same. The shooter is just 32.[152]

And while he has every morning set the time for a Shah, this Shah sets a very basic choice for his man of chronological measurement: It's the death penalty unless the watchmaker converts to Islam. Stadler does not, and he is executed for murder.

By the sword.

This is the loaded legacy of Swiss watchmaking in Persia when another Shah is toppled and a bunch of Americans are held hostage about 340 years later.

And yet Raymond Probst is considered the calm, unbiased, Helvetic presence perfect for the job of relaying messages between the captors and the American government seeking to prevent the worst. That's a wild backstory to be swept away (and true, a lot has happened in the time since the execution of clockmaker Stadler – but he perished only a few years after Cervantes and Shakespeare – were he English we'd call him an Elizabethan – which makes him feel hardly ancient or distant at all).

Swiss neutrality is a funny thing, in that it requires a person conveniently to gloss over extended periods. Or to quote the sub-headline from the former Zurich museum curator who authors an article on the episode: "Zurich watchmaker Rudolf Stadler came to a bloody end on 16 October 1637. At the age of 32, the first Court watchmaker to the Shah of Persia in Isfahan was executed by sword. His story sounds like a tale from *One Thousand and One Nights*."

Yes. That.

But what if Stadler does not shoot a man at his house late in the night

– what if his relationship with the Shah continues to flourish? What if the rectification and regulation of watches all around the palace of Isfahan encourages locals to study Stadler's techniques, apprentice to him, creates a small watchmaking cluster in the Safavid kingdom?

Do today we think of Persian watches as readily as we do Persian rugs?

If the history of horological IP were somehow truly fated, instead of a series of turns that seem wildly inadvertent and fortuitous and worthy of retelling for being so, how would it explain the demise of Persian watchmaking following a gunshot in the night and a sword to the neck?

By chance, watchmaking ebbed away in Persia. In a very slightly altered universe, Tehran attains a status like Geneva's.

* * *

The strangest thing about the watch industry is the nature of its victimology. The victims of counterfeiters are said to be the companies they've ripped off – and there's a basic legal principle in that. Fine.

The more unhinged prices get, however, the larger the incentive for even a wealthy businessperson of sound judgment to go for the very well-made, high-end clone – which probably retails from $700-$1,500 depending on quality but can go for as low as $500 if you know what you're doing.

Consider Audemars Piguet. You don't just walk in and buy a Royal Oak. The company tracks your visitation to its clubhouses before deciding whether you should be allotted a slot for one or doomed to that list which allows purchases of its less popular models – say the original Code 11.59 – but no more.

I spoke to a very high level fellow at work in the Swiss watch community with exposure to the very best products who made it quite clear: the best Chinese clone of the six-figure black ceramic Royal Oak chronograph is gorgeous, highly accurate and precise – that he has even brought one to a group of very wealthy watch collectors during a routine friendly meet-up, an event to which all the members bring pieces they'd like to discuss or display.

It was a simple test he wanted to perform. And only one of maybe two dozen experts could tell the Royal Oak wasn't quite aboveboard after legitimate scrutiny.

Now, ignore if possible, the notion that the cheapest price I can presently find online through third-party sellers for that black ceramic chrono – the 26240CE – is $120,000 (and it rises quickly from there). Focus instead on what's more ludicrous:

That even if you are perfectly willing and able to spend such dough on a timepiece that will make you a tender target for street-level thievery and mugging, you still have to visit the watch company's adult treehouses – its silly man-caves – a certain number of times just to be eligible to give AP this business.

Two wrongs don't make a right, but the greater this business drifts from transactional horology to an exclusive sort of clubbiness – I believe the current term in vogue for AP's strategy is "gatekeeping" – the more it will persuade the horologically-minded to purchase those watches whose makers want to evangelize for mechanical gearing in an impersonal, digital world – not seal that technology away in a vault...behind a velvety red cordon.

Who are the victims? Consider a story out of Osaka. A Japanese man named Takamizu Fukuhara – just a high school graduate – works for a few years in real estate, very possibly opens his own venture that fails. Then he embarks upon a path of self-reinvention, and it very much involves watches.

Now going under the alias Takamizu Kominato, he comes up with a rather innovative idea: Say you have a luxury watch stored in a safe-deposit box in a bank – your heirloom or prized piece is safe, but it's also an asset yielding you nothing as it sits there. What if there were a service that allowed customers to rent out very expensive watches for a certain sum and then evenly split the monthly loan fee with the company? So in January 2021, Kominato starts a company called Neo Reverse to operate a program for watch rentals called "Tokei Match" – which literally just means "Watch Match" (he chose English for half the name).

One customer of the service hands over a $6,400 watch – which the company rents out for about $200 per month, a full hundo of which the non-wearing customer receives now. The company quickly accumulates

1,500 watches by August 2023.

And then, the following January, at the beginning of 2024, the company announces its liquidation – but customers will receive their timepieces back, they are told. Except 860 of those watches are not returned to their owners – in fact, 20 or so are very obviously posted to an auction website.

Worse, it seems Kominato has taken a Rolex entrusted to his service in December, and, in January, flipped it to a dealer for 650,000 yen (about $4,300 at that point). And by now, it's possible he has already flipped *70 watches for $661,000.*

And by the time this news leaks out, he has already absconded to Dubai, and the UAE has no extradition treaty with Japan.

Now, the victims of this scheme are apparent – a group of 300 people whose watches have not been returned form an online group to help each other. Some are watch dealers who were looking to make some money on their inventory; others are mere individuals and less well off.

And yet, there's a commentariat that feels okay to blame them – to castigate them for taking what should be objects of great personal attachment and fascination and sentiment and trying to secure a passive income against them.

Worse, one of the victims espouses the very same idea, which is even sadder to read. From the Asahi Shimbun: "A 37-year-old office worker who lives in Yuki, Ibaraki Prefecture, regrets his decision to lend two Rolexes to Neo Reverse at the end of 2021. 'Although I lent them, they are still the precious belongings of my late father.'"[153]

And there it is – in the man's initial financial maneuver and in his subsequent sorrowful lament — all the promise and peril, earnestness and avarice, nostalgia and venality – all the deep-seated emotion and commerce a single watch, or two, can represent.

The man did nothing wrong at all. He may feel otherwise for the rest of his life.

When it comes down to it, none of this matters – the engineering prowess, the precious metal, the beautiful mathematics underpinning a gear train. We live short lives in a vast cosmic universe that preceded us and will go on without us just fine.

But the human condition is to imbue the object, almost any object, with meaning – to give us a material substitute for everything organic that can waste away – to involve ourselves in the intricacies of devices rather than resign ourselves to inevitability of the abyss.

There's nihilism…or the shimmering black bracelet of the ceramic AP. And the beauty of its 4401 chrono movement whirring around as it keeps a mechanical stopwatch record of whatever task you're timing.

The only idea in watchmaking that has never been stolen in some way or another – it's the unspoken one – the notion underpinning the entire industry:

If you're absorbed in the precise measuring of time, you can blissfully lose sight of its expanse.

In possession of watches real and fake, old and new, I'll still cry happily if I should ever have the opportunity to pass one on; the pilferage, fakery and phoniness my research has uncovered – none of it overcomes the emotion of entrusting, the idea, however false, of continuity, of bequeathal.

I wrote this book while awaiting the birth of my first child.

The watch types – the brand presidents, the CEOs, the dealers – I spoke to for this book – the more they tried to sell me on anything, the less I cared for it.

The hypnagogic sweep of the seconds hand 360 degrees at a time is a proposition seductive by itself.

If you're going to attach words to the thing, they will have to outcompete the purity of a hundred parts working together to carry us forward, the oscillations of the balance wheel nearly silent.

12

Life at 5 Hz

If this is where the emotion lives – in the sweep of the hand – then perhaps there is still something to be reclaimed. Not the stolen IP, not the lost factories, but the idea that what we make with our hands belongs first to the people who conceived it.

Across the 19th and 20th centuries, few, if any, American watchmakers or consumers stand up to Longines and its efforts to usurp American technology – to snatch away the entire US horological industry.

Of course, this is an effort not undertaken by Longines alone – *hardly* – but it is most certainly a process begun by Ernest Francillon, Longines proprietor, and his right-hand man, Jacques David, during the 1876 spying mission in Waltham, Massachusetts.

If Longines is not some singular enemy, it is most certainly that antagonist that set off the chain of events that left America bereft of horological manufacture.

But there is an entirely separate Longines scandal 70 years after the one upon which we've focused – one that I hesitate to bring up (lest this book seem a diatribe about a single brand). Yet there is Longines again – swiping from an American inventor and attempting to deny the man credit or compensation.

What is that line about history being a cycle we're condemned to repeat – unless we actually reckon with it, in all its complexity and occasional ugliness?

The year is 1947. An American named Nathan George Horwitt, a childhood émigré from Russia and a US Army veteran who served in World War I, is an ad executive at ER Squibb, a pharmaceutical firm.

But the man has talents beyond just copywriting and marketing. He attended CCNY, NYU and the Art Students League. His is a Bauhaus aesthetic – he very much believes in form following function.

So '47 – India and Pakistan become nations independent from British rule, the Yankees beat the Dodgers in the World Series, and Horwitt, pharma ad exec, tries his hand at a very different craft: the creation of a watch dial.

Only Horwitt does not follow those dials that are presently popular or that have preceded his own; rather, he strips all marking and indication stripes and indices away – he creates a spare, bare, austere watch – a black dial with but a single gold dot at the 12 o'clock position. Later, this radically minimalist dial will enter the collection of the Museum of Modern Art, earning it the sobriquet the "museum watch."

But for now, Horwitt wants to pitch commercially his design, which is unlike anything that has ever been marketed before, to the timepiece makers who dominate the market.

This goes poorly – *seemingly*. Horwitt pitches the stripped-down dial to Vacheron Constantin-LeCoultre (back then, Vacheron and what we today call Jaeger-LeCoultre were part of the same corporate structure). Horwitt is rejected and dismissed from the office. In fact, Horwitt pitches a total of 15 brands (some sources cite 13) on his jet-black-gold-dot dial. We still do not know exactly which others he approached, just the total number – and not one offers to license it.

The bosses of the watch world are simply not impressed by a dial that embodies the notion of less being so much more.

Or so they pretend.

Because we do know that one brand aware of Horwitt's work winds up *secretly* very intrigued. It is the American distribution arm of Longines – Longines-Wittnauer (which, for the sake of full disclosure, is not financially tied to the Swiss parent company in a way that would automatically suggest legal liability for what I'm about to reveal; of course, Longines-Wittnauer has an open line of communication with the Swiss parent of the watches it vends; the home base has a great deal of control).

And we also know that about a decade after Horwitt shows the major timekeeping enterprises his revolutionary dial, Longines-Wittnauer debuts a watch with what appears to be a damn similar dial (if not a copy altogether). And we know that in the interval between the pitch to companies and the Longines-Wittnauer unlicensed knock-off of this American invention, two major events occur to change the minds of those who once dismissed polymathic designer Nathan George Horwitt.

First, in 1960, the Museum of Modern Art presents the dial (and its maker) with an award and obtains the dial for its permanent collection. Next, in 1962, Horwitt enters into a legitimate licensing agreement with Movado for use of the dial (which admittedly pays him very little; in the following decade, he receives perhaps a single digit percentage of incoming monies, such that he makes $45,000 off his greatest design across 10 years).

Obviously, the success of the legal Movado dial and the imprimatur of MoMA in aggregate convince Longines-Wittnauer they have screwed up – they have ignored a designer whose genius was evident to rival tastemakers.

At which point, in the late '60s, the Longines distributor has a very obvious move – call Horwitt to headquarters and figure out a way either to buy out his Movado contract and use the original museum dial or hire the designer to make an alternate design that will still generate sales and publicity.

Ah, but Longines-Wittnauer does precisely neither: Instead, the company ignores Horwitt and simply produces a rip-off of his dial – without permission or apology. Or even shame, seemingly.

Horwitt's chess moves are limited, too: He can decide he lacks the funds and fortitude to sue a major corporation for an infringement of his patented work. Or he can take the company on, in a horological David vs Goliath battle.

He yanks out his slingshot and aims between the eyes.

Longines-Wittnauer makes a series of spurious claims in court – that the minimalist design they copied was never Horwitt's to begin with – after all, aren't sundials also stripped-down watch dials? They argue further that omitting elements of a design – the way this black-gold dial does – is an act of unprotected omission rather than safeguarded creation.

They also try to weasel out of the case via technicality – in point of fact, the copycat watches are imported in this instance by Piaget and where is Piaget located in the suit?

Ultimately, the judge has to decide whether an ordinary observer would confuse the Longines-Wittnauer version for Horwitt's and define also the nature of creativity. In fact, one could argue this is a seminal trial in the history of intellectual property: It forces dry, pedantic judges to conjure up a definition of ingenuity – to undergird the realm of beauty with legal principle.

In the end, Judge Whitman Knapp in 1975 rules that a certain spartan simplicity is itself a legitimate design choice – that this starkness is identifiable with Horwitt – is quite obviously part and parcel of his design language.

The judge calls Horwitt's dial "a concept directly contrary to the teachings of the prior art… distinctive, ornamental and aesthetically pleasing."[154]

But this encouraging line comes late in Horwitt's life, in 1975 – 28 years after he first brought his work to watchmakers for their licensing/consum ption. All that Longines-Wittnauer had to do was pay the man far earlier when he first pitched his work.

In fact, Longines does pay — but in 1973, as part of an initial legal action, in an out-of-court settlement the details of which have not been disclosed. Horwitt, for reasons I consider entirely technical and pedantic, loses his 1975 follow-up action.

But the payment speaks for itself, in my opinion. And not just mine: Movado even attempts to claim those monies paid by Longines-Wittnauer to Horwitt, since it makes use of the dial.[155]

An out-of-court settlement in 1973 for a design conceived in 1947.

Longines-Wittnauer has taken from an American creator his time — that part of life that can never be returned.

* * *

There's one more post-script here – it's the watch world – there's always another moment – this is an enterprise that counts up to 0:00 – only to begin the measuring all over again.

During the writing of the book, I acquire a 1969 Longines Ultra-Chron – the true pioneer of the commercial 5 Hz-frequency (36,000 vibrations per hour) high-beat watch.

Well, below find the photos of its gloriously preserved exterior and interior, which I duly sent to the Longines heritage department after my purchase (though not the historian there who refused to engage with his employer's 19th-century past) – along with my watch's particular numerical markers:

Movement Serial Number: 14,962,051

Case Serial Number: 270783

Movement Caliber: 431 (17 Jewels, 36,000 vph)

Longines replied that this particular watch's serial number indicated that on September 8, 1969, it invoiced its Japanese distributor for this timepiece and one other – the company *Hattori.* If that name sounds familiar, it's because that is the surname of the man who founded Seiko in 1881 – which company took his name and was not fully rebranded Seiko until 1990.

In other words, in late summer 1969, the great Japanese threat to all Swiss watchmakers – the company positioned to put them out of business with quartz technology – took control of my Longines Ultra-Chron – arguably, the most advanced Swiss automatic watch ever – so as to sell it in Japan.

My uber-Swiss watch – the piece that represented the pinnacle of mechanical achievement in the summer of the moon landing – was equally a tribute to the rise of the Japanese as arbiters of horological taste and IP. They knew from advanced movements – and whereas the Ultra-Chron beat at 36,000 vibrations per hour, Hattori/Seiko's upcoming Astron quartz piece would vibrate 29,491,200 times per hour – or to restate that: its VpH were nearly 29.5 million.

But the Astron hadn't been released yet when my Ultra-Chron reached Japan. And unlike innumerable mechanical devices from the 1960s, my

Ultra-Chron would survive another 57 years before it reached my hands without incurring any issues. It keeps time surpassingly well, its innards no less beautiful after a many-decades-long interval.

For everything the Longines historian was afraid to admit to me about his firm's legacy, here stood a piece whose survival testified to tremendous forethought, design, effort, no matter the malfeasance that set off this process.

There is much entanglement in watchmaking. My Swiss Longines beat at 5 Hz– and was tended to by Japanese enthusiasts who appreciated that impressive trait for almost six decades before I wore it, no matter that their country's pioneering watchmaking tech made 5 Hz seem an almost silly number.

That's a triumph of globalism and trade. Of cooperation. And of the enduring allure of mainsprings, balance wheels, gears and pinions.

LONGINES

LONGINES

Epilogue

This book was slated to be released several weeks before my wife's delivery of our first child. That looming-large due date. It didn't work out that way. Wheels already set in motion, the book arrived in stores days after my wife's week-35 spontaneous delivery of our cheek-heavy, squawky cherub — an adorable cashew with a full head of hair who couldn't quite work up the endurance to feed and whose inspiratory stridor sounded very much like an aching door hinge in need of WD-40. Or a macaw.

I write this epilogue on the other side:

After a month in the NICU, my firstborn child, my baby daughter, was discharged today. Time moves differently in the NICU — in every hospital room. Days and nights blend together. Minutes seem like years. Moments that should shock you fade into an obscurity born of numbness. And throughout it all, punctuating the long and arduous wait, wait, wait, there are moments of surreal, heart-moving beauty. Little miracles. Small burps and kisses. Eighty-two milliliters of milk chugged by a newborn within 30 minutes when no professional thought it possible. There are smiles that make you realize there's no place you'd rather be, even if you must escape as soon as possible. There are deep breaths that seem to make time stand still itself. Great chasms of near-silent calm amidst an otherwise clattering din.

None of those dynamics — the way our loves make time drag and rush, hurt and hearten — is the reason I wrote this latest nonfiction book, about how the Swiss watch cartel used industrial espionage and monopolistic bullying to eviscerate the US watch industry and attain a dominance not entirely deserved — by hook and by crook. I didn't write about the Chinese superclones, those watches that seem a kind of karmic revenge for the Swiss cartel's thuggish impositions because time itself is surreal — a legitimate thing that also doesn't quite exist, a force so subject to the emotions we bring to it. And yet, I now have a new child at home and all of it is clicking.

I owe everyone in the watch game, in the writing world, in my family's social circle gratitude immeasurable. Sometimes time alone isn't quite the conveyance forward it purports to be; community is.

* * *

Back home, we place our baby on her tummy on a play mat. She flirts with forward movement — enacts a grabby slither. She squirms and claws and inches her way toward the future.

Time that hasn't yet arrived, of which she has but the most basic sense.

And the purest.

Notes

PREFACE

1 Einstein, Albert and Infeld Leopold, "The Evolution of Physics," Cambridge University Press, 1938.

2 Los Angeles Evening Citizen News, "Selznick Loses 1931 Opener!" Jan. 02, 1931 ·Page 1.

3 The Wall Street Journal, "UBS Agrees to Buy Credit Suisse for More Than $3 Billion," March 19, 2023.

INTRODUCTION

4 China Economic Net, "The Rebirth of an Old Brand in Dandong, Liaoning," by Wen Jicong, July 8, 2021: http://paper.ce.cn/pc/content/202107/09/content_208351.html

AMERICA'S BIRTHDAY, 1876

5 The New York Times, "At the Antwerp Show," Sept. 6, 1885.

6 Stephen Foskett. "The Curious Case of the 'Pateck' Counterfeiter," Nov. 21, 2024, Grail-Watch. https://grail-watch.com/2024/11/21/the-curious-case-of-the-pateck-counterfei ter/

7 Stephen Foskett. "The Curious Case of the 'Pateck' Counterfeiter," Nov. 21, 2024, Grail-Watch. https://grail-watch.com/2024/11/21/the-curious-case-of-the-pateck-counterfei ter/

8 The Times, "Swiss Federal Tribunal – Extract of the Judgment," April 30, 1891.

9 Wilmington Daily Commercial, Dec. 9, 1872, page 4.

10 Horological Times, May 1978, page 14.

11 Thompson, Michael, "Corliss Centennial Engine," New England Wireless & Steam Museum; https://newsm.org/steam-e/corliss-centennial-engine/

12 Pocket Watch Database: https://pocketwatchdatabase.com/people/charles-vander-woer d

13 Swinton, John, "A Model Factory in a Model City: A Social Study of the Waltham Watch Factory," New York, 1887, page 5.

14 Chambers's Journal of Popular Literature, Arts and Sciences 4, no. 680, "Watchmaking by Machinery," 1877, page 93.

15 Favre-Perret, Edouard, "Philadelphia Exhibition 1876: Switzerland; Report Presented to the Federal High Council on the Horology Industry," 1877; translated by Richard Watkins, July 2004, page 10.

16 The Sydney Morning Herald, August 28, 1880, page 4.

17 Especially as Australian researcher Richard Watkins notes that Favre-Perret misunderstood the US technology – believing the Swiss lacking in the quantity of their machinery without considering the major deficit in relative quality – state-of-the-artness.

18 Donzé, Pierre-Yves, "Longines, From a Family Business to a Global Brand," translated by Rosamund Bandi-Tebbut, St. Imier: 2012, page 35.

19 The New York Times, "Sidelights of the Day: Show the Boss the Ad," May 9, 1953.

20 NAWCC Bulletin, "American and Swiss Watchmaking in 1876," by Jacques David, translated by Richard Watkins, June 2004.

21 Stark, Aaron, "Disrupting Time: Industrial Combat, Espionage and the Downfall of a Great American Company," New York: 2002, page 93-94.

22 Donzé, Pierre-Yves. *Longines: From a Family Business to a Global Brand.* Saint-Imier: Éditions des Longines, 2012, page 48.

23 Stark, Aaron, "Disrupting Time: Industrial Combat, Espionage and the Downfall of a Great American Company," New York: 2002, page 114.

24 Stark, Aaron, "Disrupting Time: Industrial Combat, Espionage and the Downfall of a Great American Company," New York: 2002, page 122.

25 Stark, Aaron, "Disrupting Time: Industrial Combat, Espionage and the Downfall of a Great American Company," New York: 2002, page 102.

26 Stark, Aaron, "Disrupting Time: Industrial Combat, Espionage and the Downfall of a Great American Company," New York: 2002, page 102.

27 David, Jacques. *Reports to the Intercantonal Committee of Jura Industries on the Manufacture of Watches in the United States.* Translated by Richard Watkins. Kingston, Tasmania: Richard Watkins, 2003. Originally published in 1877. Page 20.

28 David, Jacques. *Reports to the Intercantonal Committee of Jura Industries on the Manufacture of Watches in the United States.* Translated by Richard Watkins. Kingston, Tasmania: Richard Watkins, 2003. Originally published in 1877. Page 25.

29 Donzé, Pierre-Yves. *Longines: From a Family Business to a Global Brand.* Saint-Imier: Éditions des Longines, 2012, page 48.

30 Stark, Aaron, "Disrupting Time: Industrial Combat, Espionage and the Downfall of a Great American Company," New York: 2002, page 142.

31 NAWCC Bulletin, No. 350 (June 2004): pp. 294-302. Second page of article: https://thei ndex.nawcc.org/Articles/Watkins-JacquesDavid.pdf

32 Moore, Charles, "Timing a Century: History of the Waltham Watch Company," Harvard University Press, Cambridge: 1945, page 190.

33 Moore, Charles, "Timing a Century: History of the Waltham Watch Company," Harvard University Press, Cambridge: 1945, page 77.

34 Herald Statesman, "Axlers Report Shah's Visit at Sans Souci," Jan. 21, 1955, page 13.

35 Boston Globe, "N.Y. Syndicate Buys Waltham Watch Company," Oct. 9, 1956, page 1 and page 3.

36 The New York Times, "Harold Danzig, new president of Waltham Watch, pins his hopes on a new product," March 5, 1978, section F, page 1.

37 The New York Times, "Inah Canabarro Lucas, 116, the World's Oldest Person," May 4, 2025, section 1, page 34.

38 The Boston Globe, Jun 17, 1992, page 63.

THE ORIGIN OF EVERYTHING: FRANCE, 1800

39 Stendhal, "Rome, Naples and Florence," Paris: 1854 (Richard N. Coe translation), page 150.

40 https://escapementmagazine.com/articles/abraham-louis-breguet/#:~:text=Image%20%E2%80%93%20Breguet%20hour%20and%20minute,another%20Parisian%20watchmaker%2C%20Xavier%20Gide.&text=Gide%20seems%20to%20have%20been,and%20Daniels%20describes%20him%20as%20%E2%80%9C%E2%80%A6

41 https://www.beyer-ch.com/en/breguet/#:~:text=The%20name%20Breguet%20is%20part,father%20of%20modern%2Dday%20watchmaking.

42 Canterbury Christ Church University: "David Lionel Solomons – master of electricity and theatricality": https://web.archive.org/web/20110720144338/http://www.canterbury.ac.uk/Solomons-museum/david-lionel-Solomons.asp

43 https://jch.history.ox.ac.uk/article/young-photographer-broomhill

44 Vanity Fair, "David Lionel Solomons – Master of electricity and theatricality," June 17, 1908.

45 The New York Times, "Marie Antoinette's Watch, Which She Never Saw, Is on Display," Jan. 29, 2025.

46 Daniels, George and Markarian, Ohannes, "Watches & Clocks in the Sir David Solomons Collection," Page 62.

47 The Telegraph, "Breguet No106 Watch: The queen, her watch and the master burglar," by Alex Krista, April 15, 2009.

48 Toronto Star, "Museum hit in biggest Israeli heist," April 18, 1983, page 8.

49 The New York Times, "Art: Clocks at the Frick," Jan. 7, 1983, Section C, page 23.

50 Haaretz, "The Great Watch Robbery: Naaman Diller – the rejected kibbutznik who continues to surprise even after his death," Oct. 29, 2008.

51 Maariv, "January 11, 1980, page 65.

52 Haaretz, "The Great Watch Robbery: Naaman Diller – the rejected kibbutznik who continues to surprise even after his death," Oct. 29, 2008.

53 John Biggs, "The true story of Marie Antoinette's amazing $11 million watch," Medium, Apr 15, 2016: https://johnbiggs.medium.com/the-true-story-of-marie-antoinette-s-amazing-11-million-watch-cc6189337105

54 Biggs, John, "Marie Antoinette's Watch: Adultery, Larceny & Perpetual Motion," Ray Bridge Press: 2015, Page 197.

55 "In Depth: Breguet No. 160 Montre d'Or," by Carlos Torres, for SJX, March, 23, 2026: https://watchesbysjx.com/2026/03/breguet-no-160.html

PRE-WESTERNIZED JAPAN, 1860

56 "Watches, Watch Movements, And Watch Parts": Report to the President on Investigation No. 337-19," TC Publication 177, Washington, DC, June 1966.

57 The Province (Vancouver, British Columbia), "Swiss Watchmakers to Regulate Output," June 5, 1927, page 32.

58 Donzé, Pierre-Yves, "The Business of Time," Manchester University Press: Page 66.

59 The Argus-Farmer (Lewistown, Montana), "Co-operation ot Solve Swiss Watch Makers' Problems," June 30, 1927, page 23.

60 Donzé, Pierre-Yves, "The Business of Time," Manchester University Press: Page 66.

THE SWISS MOB AND AMERICA, 1936

61 The New York Times, "US Signs Treaty with Switzerland," January 10, 1936, page 14.

62 **Foreign Relations of the United States, Diplomatic Papers, 1936, Europe, Volume II** 611.5431/498a: Telegram

63 The New York Times, "Watches From Switzerland," Aug. 17, 1952, Review page 8.

64 The New York Times, "TRUMAN IS TOASTED BY GRATEFUL SWISS," Aug. 16, 1952, Page 1.

65 Time Magazine, "The Watch Tariff," July 12, 1954.

66 Time, "Governenment Alarm Over Watches," Nov. 1, 1954.

67 Boillat, Johann, "Les véritables maîtres du temps : le cartel horloger suisse (1919-1941)," Éditions Alphil, 2013, page 610.

68 Donzé, Pierre-Yves, ""History of the Swiss Watch Industry: From Jacques David to Nicolas Hayek" (first published in English by Peter Lang in 2011/2012), page 2.

69 Time, "Governenment Alarm Over Watches," Nov. 1, 1954.

70 Time, "Governenment Alarm Over Watches," Nov. 1, 1954.

71 The Progress-Index, "A Decision on Watches," Jan. 16, 1965, page 4.

72 The New York Times, "24 U.S., Swiss Concerns Accused Of Rigging Jeweled Watch Trade," October 20, 1954, BUSINESS FINANCIAL, Page 41.

73 Donzé, Pierre-Yves, "The Business of Time," page 80.

74 Donzé, Pierre-Yves, "The Business of Time," page 80.

75 **Memorandum Of United States In Response to Motion of Rolex Watch U.S.A., Inc. For Order Terminating Final Judgment,** February 28, 2006: https://www.justice.gov/atr/case-document/memorandum-united-states-response-motion-rolex-watch-usa-inc-order-terminating

76 June 1966 US Tariff Commission Report, page 3.

77 June 1966 US Tariff Commission Report, page 8.

78 June 1966 US Tariff Commission Report, page 9.

79 June 1966 US Tariff Commission Report, page 10.

80 Lancaster New Era, "Swiss Attorney Answers Charge By Sinkler," July 21, 1965, page 30.

81 Time Magazine, "The Watch Tariff," July 12, 1954.

82 Donzé, Pierre-Yves, "The Business of Time," page 94.

ASIA GOES ITS OWN WAY: JAPAN, 1930; HONG KONG, 1973

83 Hirano Mitsuo, *Tokeiō Hattori Kintarō* [], Ichigyō Ichinin Den (Tokyo: Jiji Tsūshinsha, 1972), pages 1-48.

84 Hirano Mitsuo, *Seikōsha shiwa* [Historical Account of Seikōsha] (Tokyo: Seikōsha, 1968)

85 Mainichi Shimbun, "Expanding its presence into the heart of Ginza," Nov. 11, 2023: https://mainichi.jp/articles/20231106/oog/00m/020/003000c

86 https://museum.seiko.co.jp/en/knowledge/relation_08/

87 https://museum.seiko.co.jp/seiko_history/founder/founder/006/

88 Hirano Mitsuo, *Seikōsha shiwa* [Historical Account of Seikōsha] (Tokyo: Seikōsha, 1968).

89 The Financial Post, "Seiko Makes Jeweled Lever Pay Off," August 22, 1970, page 32.

90

Dodis- Diplomatic Documents of Switzerland: dodis.ch/32775

3.9.1968 (Tuesday)

Uhrenabkommen UdSSR

Memo (No)

Das Rahmenabkommen zwischen der schweizerischen Uhrenkammer mit der UdSSR wurde noch nicht unterzeichnet. Ebenso wie die Mission des Vororts nach Moskau wurde dieses Unterfangen wegen der Ereignisse in der Tschechoslowakei vorübergehend auf Eis gelegt.

File reference: 3200.13

91

Dodis- Diplomatic Documents of Switzerland: https://dodis.ch/32775

3.9.1968 (Tuesday)

Uhrenabkommen UdSSR

Memo (No)

Das Rahmenabkommen zwischen der schweizerischen Uhrenkammer mit der UdSSR wurde noch nicht unterzeichnet. Ebenso wie die Mission des Vororts nach Moskau wurde dieses Unterfangen wegen der Ereignisse in der Tschechoslowakei vorübergehend auf Eis gelegt.

File reference: 3200.13

92 3.1969, **Report on meetings with industry and factory visits**

Discussions with representatives of the chemical, watchmaking, and machinery industries revealed a very keen interest in trade with the USSR. Detailed list of the results of an inquiry by the Industrial Council of the Swiss Embassy in Moscow. dodis.ch/32694

Page 14

93 **Dodis- Diplomatic Documents of Switzerland:** dodis.ch/33591

5.11.1969 (Wednesday)

Argentinien - Uhren/Wein

Letter (L)

Die Schweiz möchte das Junktim zwischen Zollsenkung für den Import von Uhren und die Anerkennung eines zusätzlichen Weinkontingents an Argentinien aufheben, doch hält Bern in den Verhandlungen mit Buenos Aires momentan keine Trümpfe in der Hand.

File references: 562.12 • Arg.873.1 • 551.31 • Arg.841.8.AVA • Arg.861.7.Allg

94

Dodis- Diplomatic Documents of Switzerland: dodis.ch/35985

9.9.1970 (Wednesday)

Notiz über die Besprechung mit den Herren Präsident Bauer, Direktor Retornaz und Schütz von der F.H. vom 21.8.1970 bei Herrn Botschafter Probst (Vorsitz)

Memo (No) • confidential

Orientierung über Probleme der schweizerischen Uhrenindustrie in Lateinamerika. Der von Uhrenimporteuren befürchteten japanischen Konkurrenz soll entgegengehalten werden. Hierzu werden verschiedene Vorgehensweisen in Mexiko, Argentinien, Uruguay, Brasilien und Peru verglichen.

File reference: Bras.870

95 Un homme et son empreinte: Hommage à Gérard Bauer pour ses 80 ans, 8 juin 1987. Edited by Etienne Jeanneret under the direction of Jean Cavadini. Hauterive, Switzerland: Editions Gilles Attinger, 1987.ISBN 2-88256-027-3. 213 pp.(Festschrift; ~24 contributors. Volume formally presented to Bauer at the Musée d'histoire naturelle de Neuchâtel, 30 October 1987.) Pages 151-152.

96 Time, "Business & Finance: Electric Watch," February 29, 1932.

97 The New York Times, "SIMPLIFIES TYPEWRITER; Swiss Says Small Model of His

Electric Machine Will Cost $2," March 25, 1932, page 7.

98 Pueblo Chieftain, "Howard Will Speak at Welfare Session," Oct. 11, 1946, page 8.

WAS IT ALL ABOUT MARX FROM THE START? OR, A DEEPER LOOK AT LONGINES

99 Donzé, Pierre-Yves, "Longines, From a Family Business to a Global Brand," Editions Des Longines, page 30.

100 Eitel, Florian, "Le Vallon horloger et ses anarchists," Intervalles, No. 123, 2022.

101 Garfield, Simon, "Timekeepers," Canongate: Edinburgh, page 46.

102 Eitel, Florian, "Anarchistische Uhrmacher in der Schweiz," pages 555–556. For the specific part of the Longines payroll in which the overlapping worker are found, see Serie E (Personnel), Sous-serie 122.3 (Livrets d'ouveriers).

HONG KONG BY WAY OF THE GOLDEN GATE, 1966

103 Donzé, Yves, "The Business of Time," Manchester University Press, Page 111.

104 Economic History of Hong Kong, Catherine R. Schenk, University of Glasgow: https://eh.net/encyclopedia/economic-history-of-hong-kong/

105 Time Magazine, "HONG KONG: The Fragrant Harbor," Nov 21, 1960.

106 Time Magazine, "HONG KONG: The Fragrant Harbor," Nov 21, 1960.

107 "The Watch Industries in Switzerland, Japan and the United States," World Watch Case (Calcutta: Indian Institute of Management, n.d.), page 3.

108 "The Watch Industries in Switzerland, Japan and the United States," World Watch Case (Calcutta: Indian Institute of Management, n.d.), page 4.

109 **Dodis- Diplomatic Documents of Switzerland:** dodis.ch/32551

15.6.1967 (Thursday)

Uhrenschalen aus Hong Kong

Memo (No)

Der Begriff "Swiss made" ist im Uhrensektor unklar, was in Hongkong zu Problemen führt. Dort werden Imitationen als schweizerische Modelle verkauft.

File reference: Hong.865.2

110 Pierre-Yves Donzé, The Business of Time: A Global History of the Watch Industry, Studies in Design and Material Culture (Manchester: Manchester University Press, 2020), page 87.

111 Donzé, Pierre-Yves, "The Business of Time," page 87.

112 *L'Impartial* (La Chaux-de-Fonds) « **Politique d'abandon de la montre suisse** » 16 November 1966, page 9: e-newspaperarchives.ch — IMP19661116-01.2.45

113

Dodis- Diplomatic Documents of Switzerland: dodis.ch/32551

15.6.1967 (Thursday)

Uhrenschalen aus Hong Kong

Memo (No)

Der Begriff "Swiss made" ist im Uhrensektor unklar, was in Hongkong zu Problemen führt. Dort werden Imitationen als schweizerische Modelle verkauft.

File reference: Hong.865.2

114 Entretien du 18.4.1973 entre MM. Bauer, Retornaz, Ponnaz, Hool, Thomas (FH), Benoit (Interpol) et MM. Probst, Töndury et Grossenbacher: dodis.ch/38695

115 The entire meeting, transcribed in German, is kept for posterity online via electronic scan of an old typewritten document by a Swiss Government archive called Dodis— which stands for "Diplomatic Documents of Switzerland" — here: https://dodis.ch/38709

Dodis labels the meeting thusly:

24.9.1973 (Monday)

Uhrenprobleme in Hong Kong

Zusammenfassung der Unterredung zwischen der Schweizer und der Hong Konger Delegation auf dem Departement of Commerce & Industry mit enttäuschendem Ausgang für die Schweiz.

116 Conversation on February 22, 1973, with Mr. Dominique Thomas, Director of External Affairs at FH, and Mr. Max Hool, Administrative Director and Head of the Counterfeiting Department at FH, in the office of Ambassador Probst. dodis.ch/38693

THE GREAT OPENING, MAINLAND CHINA, 1979

117 The State Journal-Register, "China Speeds Modernization," Dec. 14, 1979, page 49.

118 "Chinese Wooing Houston Business," The Houston Post, April 28, 1984, page 1C.

119 Derby Evening Telegraph (Derby, England), "Trade: Why China Woos the West," Oct. 31, 1979, page 41.

120 "Chinese Riot Linked to Escape Bid," Baltimore Sun, June 23, 1979, page A2.

121 "A Peaceful Town in China Transformed by Industry," The San Francisco Examiner, Dec. 5 1979, page 69.

122 "Chinese woo local business with profit lure," The Houston Post, April 28, 1984, page C2.

123 "Chinese woo local business with profit lure," The Houston Post, April 28, 1984, page C2.

124 "China: The Next Oil Giant?" Newsweek, Oct 27, 1975, page 40.

125 "China Offshore Oil 'Boom' a Bust, Western Firms Find," Los Angeles Times, Nov. 17, 1985.

126 Knusel, Ariane, "China's European Headquarters," University of Cambridge Press: 2024, page 121.

127 Knusel, Ariane, "China's European Headquarters," University of Cambridge Press: 2024, page 134.

128 Knusel, Ariane, "China's European Headquarters," University of Cambridge Press: 2024, page 134.

129 Knusel, Ariane, "China's European Headquarters," University of Cambridge Press: 2024, page 134.

130 Knusel, Ariane, "China's European Headquarters," University of Cambridge Press: 2024, page 128.

131 Donzé, Pierre-Yves, "The Business of Time," page 112.

132 Knusel, Ariane, "China's European Headquarters," University of Cambridge Press: 2024, page 135.

133 Knusel, Ariane, "China's European Headquarters," University of Cambridge Press: 2024, page 135.

134 Journal of Contemporary History, "'No Ski Shoes, Chocolate or Hairsprays': The (Mis)Adventures of European Companies in China, 1978-88," by Ariane Knusel, Volume 60, Issue 3, March 13, 2025: https://journals.sagepub.com/doi/10.1177/0022009425132 3798

135 Journal of Contemporary History, "'No Ski Shoes, Chocolate or Hairsprays': The (Mis)Adventures of European Companies in China, 1978-88," by Ariane Knusel, Volume 60, Issue 3, March 13, 2025: https://journals.sagepub.com/doi/10.1177/0022009425132 3798

136 **Dodis- Diplomatic Documents of Switzerland:** dodis.ch/32551

15.6.1967 (Thursday)

Uhrenschalen aus Hong Kong

Memo (No)

Der Begriff "Swiss made" ist im Uhrensektor unklar, was in Hongkong zu Problemen führt. Dort werden Imitationen als schweizerische Modelle verkauft.

File reference: Hong.865.2

137 Donzé, Pierre-Yves, "The Business of Time," page 113.

138 A Blog to Watch, "Earnshaw Beagle Watch Review," by Ariel Adams, July 29, 2013: https://www.ablogtowatch.com/earnshaw-beagle-watch-review/

PEARL RIVER DELTA, DANDONG, PADUA, 2015

139 Annals of Medical History, ""The Multiple Personality of Dr. Guy Patin,"

140 Vanity Fair, "La chute de Julien V., le prince de la fausse montre de luxe," June 23, 2024.

141 The following account of Peacock's involvement with Noob has been confirmed for me by sources involved in this investigation.

142 Vanity Fair, "La chute de Julien V., le prince de la fausse montre de luxe," June 23, 2024.

143 Revue FH, "Dismantling the' *Prestige Replica'* Network," Number 8, June 6, 2024

144 The entire Vincent element of the Noob raid is covered splendidly and in depth by Passion Horologere (https://passion-horlogere.com/contrefacons-de-montres-la-plus-grande-af

faire-jugee/) , Le Parisien (https://www.leparisien.fr/faits-divers/rolex-patek-philippe-le-plus-important-faussaire-de-montres-de-luxe-compte-les-heures-en-prison-29-08-2023-DAIBIM5M6JHGFMURVES2IV27TQ.php#:~:text=En%20pr%C3%A8s%20de%20quatre%20ann%C3%A9es,r%C3%A9v%C3%A9l%C3%A9%20Le%20Parisien%2C%20Florian%20R.) and by Vanity Fair (https://www.vanityfair.fr/article/la-chute-de-julien-v-le-prince-de-la-fausse-montre-de-luxe)

145 Here's what my Western source said when we discussed confirmation of this tale: *"The discussion yesterday with one of our informants... he said, well, what I can tell you is that they did receive a strong letter from Rolex lawyers. So he was aware, I didn't say anything. He was aware of that."*

THE BODY KEEPS THE SCORE

146 Neue Zürcher Zeitung Folio, "Der talentierte Monsieur Jaquet," by Lukas Egli, April 1, 2009.

147 "Swiss Police Tick Off Fake Watch Makers," Agence France Press, Oct. 13, 2003.

148 Neue Zürcher Zeitung Folio, "Der talentierte Monsieur Jaquet," by Lukas Egli, April 1, 2009.

149 Neue Zürcher Zeitung, "Der talentierte Monsieur Jaquet," by Lukas Egli, April 1, 2009.

150 "Affaire Jaquet": jusqu'a neu fans de reclusion," Swissinfo.ch, Nov. 3, 2008.

ALL OF US

151 "New York. Failure of Tea Dealers—The Murder of Delia Corcoran—A Boy Murder—Immigrants Arrived—The Watch Forgery Cases," July 30, 1873, page 1.

152 CLOCKS in Encyclopedia Iranica, by Floor, Willem M. Last Updated June 25, 2013: https://www.iranicaonline.org/articles/clocks-pers/
Print Detail: Vol. V, Fasc. 7, pp. 713-718, Published December 15, 1992.

153 The Asahi Shimbun, "Owners of luxury watches fear scam in 'sharing economy' service," by Kaede Sano, February 12, 2024: https://www.asahi.com/ajw/articles/15162104

LIFE AT 5 HZ

154 The case: *Horwitt v. Longines-Wittnauer Watch Co., Inc.*, 388 F. Supp. 1257 (S.D.N.Y. 1975)
Court: U.S. District Court, Southern District of New York
Judge: Whitman Knapp
Tried: October 30–31, 1974
Decided: January 22, 1975

155 **Horwitt v. Movado Watch Agency, Inc., 364 F. Supp. 687 (S.D.N.Y. 1973)**
Full Name: Horwitt v. Movado Watch Agency, Inc.
Citation: 364 F. Supp. 687
Docket Number: 71 Civ. 5537
Date: September 13, 1973

www.ingramcontent.com/pod-product-compliance
Lightning Source LLC
LaVergne TN
LVHW100305210926
848480LV00003B/1026
* 9 7 9 8 2 3 4 0 5 4 6 4 7 *